Safety and health in the use of chemicals at work

A training manual

Safety and health in the use of chemicals at work

A training manual

Abu Bakar Che Man and David Gold

An ILO contribution to the
International Programme on Chemical Safety
(a collaborative programme of the United Nations Environment Programme,
the International Labour Organization
and the World Health Organization)

International Labour Office Geneva

First published 1993
Second impression 1998

Bakar Che Man, A.; Gold, D.
Safety and health in the use of chemicals at work: A training manual
Geneva, International Labour Office, 1993

/Trainers manual/,/Guide/,/Occupational safety/,/Occupational health/, /Chemicals/. 13.04.2
ISBN 92-2-106470-0

ILO Cataloguing in Publication Data

ILO publications can be obtained through major booksellers or ILO local offices in many countries, or direct from ILO Publications, International Labour Office, CH-1211 Geneva 22, Switzerland. Catalogues or lists of new publications are available free of charge from the above address.

Printed by the International Labour Office, Geneva, Switzerland

Preface

The International Labour Conference adopted, in June 1990, the Chemicals Convention (No. 170), and Recommendation (No. 177), with a view to reducing the incidence of chemically induced illness and injuries at work.[1]

Over the past decade there has been a vast increase in the use of chemicals, and this trend will continue as chemicals have a direct impact on the improved quality of life. There are, however, many risks associated with the unsafe use of chemicals at work. Therefore safety and health in the use of chemicals offers a challenge to governments, and to employers and workers and their representatives. By strengthening tripartism and encouraging comprehensive efforts by all parties concerned, we can reduce the potential for both illness and injury and learn to work safely with chemicals.

This training manual addresses all aspects of dealing with chemicals including production, storage, use, transport and disposal. It provides an approach to problems and solutions concerning safety and health in the use of chemicals at work. Workers and their representatives will find information that will help them understand how chemicals can affect them and what measures can be taken to afford protection. Information is also provided on preventing fires and explosions. Managers and supervisors will find guidance on preventive activities and the management of a safety programme for the use of chemicals. Government officials and trainers will discover that the manual is in a format that can easily be used as the basis for a one-week training course for managers, supervisors and workers' representatives.

The manual is divided into the following chapters: the health effects of chemicals; fire and explosion hazards; methods of prevention; emergency procedures; and management of safety in the use of chemicals. These five units can be taught as a five-day course or can be presented individually, depending on the needs of the intended audience.

The ILO appreciates the direct assistance given by the Factories and Machinery Department of the Ministry of Labour of Malaysia, which provided a venue and an audience to pilot the material for this manual. Thanks are also due to Ms. Noha Karanuh, the graphic artist who illustrated the manual.

K. Kogi,
Director,
Working Conditions and
Environment Department

[1] Annex 2 gives the text of the Convention and Recommendation.

Contents

Tables

Figures

1. Introduction

Fifty years ago only 1 million tons of chemicals were produced annually. Little was known, and little was done, about the hazards associated with chemicals and chemical processes. Today over 400 million tons of chemicals are produced annually, and of the 5-7 million known chemical substances over 80,000 are marketed. Over 1,000 new chemicals are produced each year. It is estimated that 5,000-10,000 commercial chemicals are hazardous, of which 150-200 are considered likely to cause cancer.

Chemicals have improved the quality of life. Agrochemicals in the form of pesticides and fertilizers have greatly increased food production. Chemotherapeutic drugs have contributed to the treatment of cancer and new drugs are constantly entering the market for the treatment of heart disease. Carbon fibres are widely used in the manufacture of new lightweight materials, while ceramic fibres are used as insulation materials and have frequently been used as a replacement for asbestos. Acrylic adhesives, superglue and environmentally safe biodegradable plastics are other examples of the important contribution of chemicals to daily life.

Today, in virtually every workplace, workers are exposed to chemicals. Chemicals such as solvents are used for cleaning and degreasing, mixing paints and varnishes, and diluting concentrated compounds and mixtures. Chemicals in the solid state may be transformed into powders or dust particles during the manufacturing process and may rest airborne for long periods of time. Gases and vapours are employed in industrial operations such as welding and refrigeration, or in a variety of chemical processes. Gases are also used as anaesthetic agents in hospitals. Laboratories in schools, universities, research institutions, government agencies and private enterprises make use of a variety of chemicals in both large and small quantities. In agriculture, workers may be exposed to chemical-based products such as fertilizers, pesticides and herbicides. Many chemical-based pesticides are used for controlling insect-borne diseases such as malaria. Even in today's modern office, one may find a variety of different chemicals.

However, certain chemical substances can both harm and kill. Chemicals alone in concentration, or when mixed with other chemical substances, can cause injury, disease or death. Their misuse may also result in fires and explosions. It is imperative that everyone who could potentially come into contact with chemicals should know and understand the risks, and the methods available for reducing them.

This manual covers the following topics:

- health risks resulting from exposure to chemical hazards at work;
- chemical fire and explosion hazards;
- basic principles of prevention;

Figure 1.

One of the basic concepts of this manual is that there must be adequate information about every chemical in the workplace

- chemical emergency procedures;
- the management of a chemical control programme.

We hope that the information given in this manual will be of value to all those engaged in promoting or practising the safe use of chemicals at work: managers, supervisors and workers' representatives, health and safety officers and trainers. It is written in straightforward language and assumes a minimum of technical knowledge. It is thus particularly suitable for use on training courses.

The emphasis has been placed throughout on practical guidance on the safety precautions to be taken when using chemicals at the workplace. The discussion and activities included at the end of sections and chapters may be used equally by trainers in group exercises or by individual readers in self-evaluation. Suggestions for further reading are given at the end of each chapter. Finally, a comprehensive checklist for safety and health in the use of chemicals at work is given in Annex 1.

Throughout the manual three concepts will be frequently referred to:

- there must be adequate information and understanding about every chemical in the workplace in the form of labels and chemical safety data sheets (figure 1);
- there must be a clearly established policy on the safe use of chemicals in the workplace, providing the framework for both organizational and operational control measures;
- the management of chemical hazards is a task requiring the participation of both employers and workers. It starts before the receipt of the chemical and continues without interruption, ending when the chemical is neutralized or destroyed.

With these three concepts in mind we shall address the safe use of chemicals at work.

Remember:

An up-to-date label and a current chemical safety data sheet are the best sources of information on chemicals at the workplace.

2. Health hazards due to chemical exposure

A great deal of attention in recent years has been focused on the effects of exposure to chemicals on the health of workers. Many chemicals which were once regarded as safe have been found to be associated with diseases ranging from mild skin rashes to chronic health impairment and fatal cancers. Although much has been learned about chemical toxicity from the study of diseases in laboratories and elsewhere, there are far too many chemicals used in the workplace today whose harmful effects are still unknown. It is therefore essential to treat *all* chemicals with care.

This chapter will explain how chemicals affect workers' health. It will enable users to recognize a potentially hazardous situation, respond accordingly, and protect themselves and others. It will also allow the necessary steps to be taken to remedy the situation.

2.1. Definitions

To enable users to understand the health hazards of chemicals, certain terms which are used frequently in chemical safety are defined here:

Chemical — chemical elements and compounds, and mixtures of them, whether natural or synthetic.

Poisoning — normally the human body is able to cope with a variety of substances, within certain limits. Poisoning occurs when these limits are exceeded and the body is unable to deal with a substance (by digestion, absorption or excretion).

Toxicity — the inherent potential of a chemical substance to cause poisoning. The toxicity of chemicals varies widely. For example, a few drops of a given chemical will cause death while other chemicals will produce the same effect only after a large quantity has been consumed.

Hazard — a potential to cause danger to life, health, property or the environment.

Chemical hazard — any chemical that has been classified as hazardous or for which relevant information exists to indicate that it is hazardous.

Risk — the measured probability of an event to cause danger to life, health, property or the environment.

Airborne dust — refers to the suspension of solid particles in the air. These dust particles are generated by handling, grinding, drilling and crushing operations where solid materials are broken down. The size of these particles ranges from being visible to the naked eye (i.e. greater than one-twentieth of a millimetre in diameter) to being invisible. Invisible dust will remain airborne for a long period of time and is dangerous because of its ability to penetrate deeply into the lungs.

Vapour — the gaseous form of a liquid at room temperature and pressure. Liquids emit vapours, the quantity depending on their volatility. Substances with a low boiling-point are more volatile than those with a higher one.

Mist — the dispersion of liquid particles in the air. Mists are normally generated by processes such as electroplating and spraying where liquids are sprayed, splashed or foamed into fine particles.

Fumes — solid particles formed from condensation of substances from the vapour state. Fumes are normally associated with molten metals where the vapours from the metal are condensed into solid particles in the space above the molten metal. The size of the particles are in the range visible to the naked eye.

Gas — a substance, such as oxygen, nitrogen or carbon dioxide, which is in the gaseous state at room temperature and pressure.

Acute effect — the effect caused by a single short-term exposure (usually not more than one work shift) to a high amount or concentration of a substance.

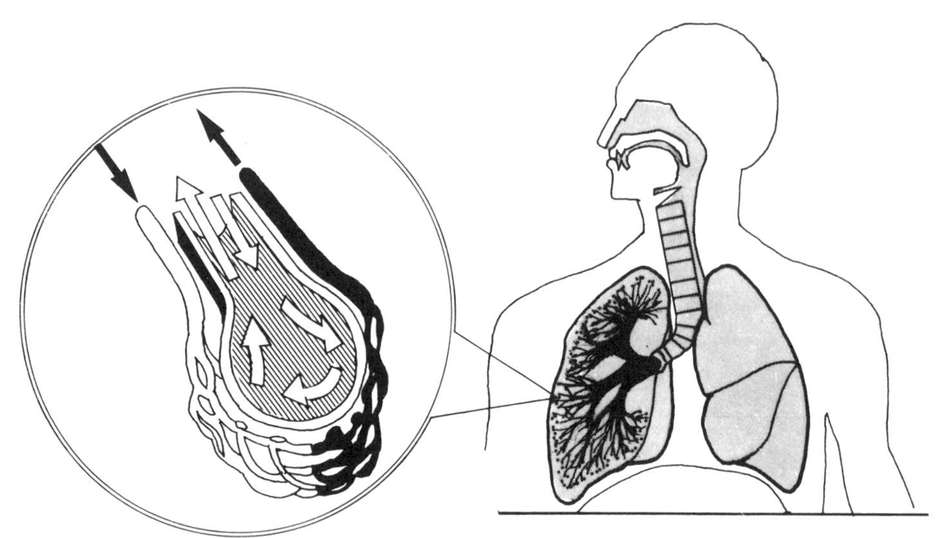

Figure 2.

Air travels into the lungs and circulates there in tiny air sacs where oxygen and carbon dioxide are exchanged

Chronic effect — the effect caused by repeated exposure to a chemical over a long period of time. The effect may be felt only after many years of exposure. Both acute and chronic effects can be reversible after the termination of the exposure and appropriate treatment, or they may result in long-lasting, irreversible conditions.

Chemical safety data sheet — a document containing essential information for users regarding the properties of chemicals classified as hazardous and methods of using them safely, including their identity, supplier, classification, hazards, safety precautions and emergency procedures.

2.2. Factors contributing to hazardous situations

Many factors can influence the intensity of hazards associated with chemicals in the workplace. These include toxicity, the physical properties of the substances, work practices, the nature of the exposure, combined exposures, the routes of entry, and the susceptibility of workers. It is important to understand how these factors, in combination, contribute to hazardous situations.

2.2.1. Routes of entry

Chemicals can enter the body in three ways. In the workplace, the inhalation of gases, vapours

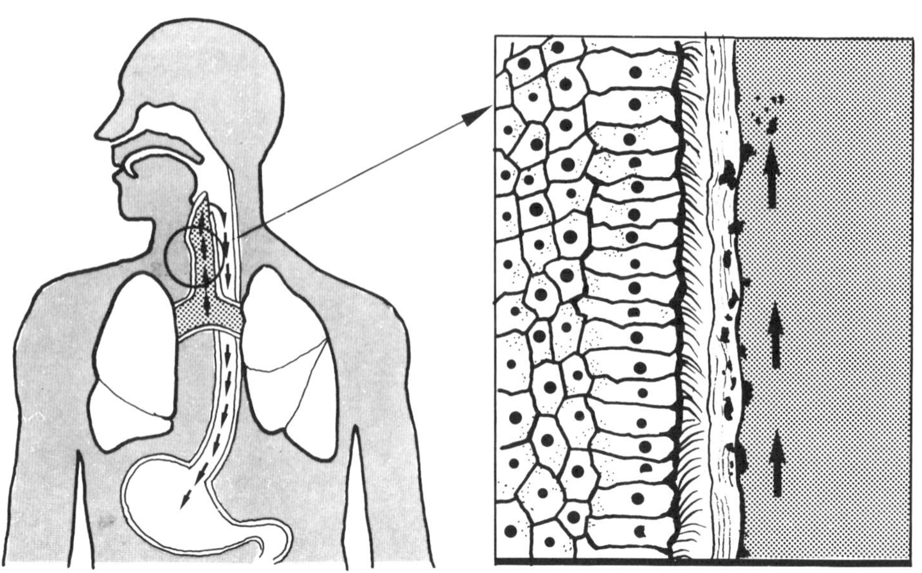

Figure 3.

Tiny hair-like structures in the air passageways help to clear them of foreign materials

or airborne particles and absorption through the lungs is the most important route of entry. However, a number of chemicals, particularly liquids, can be absorbed through intact skin when coming into direct contact with it. The ingestion of poisons through the mouth is common where personal hygiene is poor.

2.2.1.1. Inhalation

In industry, inhalation is the most significant route of entry. The respiratory system represents an efficient entry point for chemicals. With a total surface area of the lungs of 90 square metres in a healthy adult, a worker performing a moderate task inhales about 8.5 cubic metres of air in the course of an eight-hour shift.

The respiratory system consists essentially of the upper respiratory tract (nose, mouth, throat), the air passageways (trachea, bronchi, bronchioles, alveolar ducts) and the gas exchange area (alveoli) where oxygen from the air diffuses into the blood and carbon dioxide from the blood diffuses into the air (figure 2).

The air passageways are lined with tiny hair-like structures (cilia). These structures are part of the clearing mechanism of the lungs which causes foreign particles deposited on the surfaces of the respiratory passages within the lungs to be carried by mucus towards the throat (figure 3). It is estimated that 2 litres of mucus flow to the throat each day.

During breathing, airborne chemicals enter the nostrils or mouth, pass through the air passageways and finally reach the gas exchange area where they are either deposited or pass through the wall of the area into the bloodstream.

Certain substances irritate the mucous membrane of the upper respiratory tract and respiratory passages within the lungs. This irritation may serve as a warning of the presence of chemicals. However, certain gases or vapours do not have this effect. Unnoticed by the workers, they penetrate deeply into the lungs causing lung injury, or become transported in the bloodstream (figure 4).

The entry of dust particles into the body depends on their size and solubility. Only small particles (less than seven thousandths of a millimetre in diameter) will be able to reach the gas exchange area. This respirable dust (which reaches the gas exchange area) will either be deposited there or diffused into the bloodstream, depending on the solubility of the chemicals. Insoluble dust particles are mostly eliminated by the clearing mechanisms of the lungs. The larger

Figure 4.

The left-hand drawing shows an open degreasing tank giving off harmful vapours. The right-hand drawing shows a tank with a lid, thereby reducing the risk of the worker coming into contact with the vapours

dust particles are filtered by the hairs of the nostrils or deposited along the path from the nose to the air passageways. They will eventually be transported to the throat where they will be either swallowed, or spat or coughed out.

Remember:
Extreme care must be taken because chemicals in the form of vapour, fumes, dust or gas can easily enter the body through breathing.

Discussion questions:

1. List chemical substances in your workplace that may enter the body through inhalation.
2. Describe the precautionary measures taken for these substances.

2.2.1.2. Ingestion

Ingestion is another way in which chemical substances can enter the body. Entry via ingestion is possible when workers eat or smoke with contaminated hands or eat their meals at their workstation where food and drink may be contaminated by vapours in the air (figure 5).

A second way in which chemical substances are ingested is when inhaled particles are transported to the throat by the air passageways into the lungs, and swallowed.

The digestive system consists of the oesophagus, the stomach, and the small and large intestine. Absorption of food and other substances, including ingested hazardous chemicals, occurs primarily in the small intestine.

Remember:
If you eat or drink at your workstation, you may be introducing hazardous chemicals into your digestive system because the substance may coat the food or eating utensils.

Discussion questions:

1. List any chemicals in your workplace that may enter the body through ingestion.
2. What simple measures can you take to avoid ingesting chemicals into your body?

Figure 5.

It is dangerous to consume food and drink, or to smoke, at a workstation where chemicals are used. The food or drink may be contaminated by dirty hands or even vapours in the air. Cigarettes may be a fire or explosion hazard

2.2.1.3. Skin absorption

Absorption through the skin constitutes another route of entry. The thickness of the skin, together with its natural covering of sweat and grease, provide some protection against chemical exposure.

The solubility of chemicals (such as organic solvents and phenol) in fats enables their absorption through the skin. If the skin is damaged by cuts or abrasions, or is diseased, the chemical would be absorbed into the body even quicker.

Discussion questions:

1. List any chemicals in your workplace that can be absorbed through the skin.
2. Explain the measures taken to avoid skin contact with these chemicals.

2.2.2. Concentration and type of exposure

Chemical substances that enter the body through inhalation, ingestion or skin absorption are transported by the bloodstream. Some of these chemicals will be stored in tissues or organs, with very little being excreted. Some will be changed into other substances which are more soluble and will leave the body through the urine. Others will be excreted unchanged via breathing or urination. These substances may cause damage to internal organs. The breaking down and detoxification of certain substances (normally occurring in the liver) may produce by-products or new substances which may be more harmful than the original ones. The damage done by a chemical to a specific organ depends in principle on the amount (dose) absorbed. In the case of inhalation, the dose depends mainly on the concentration of the substance in the air and the duration of the exposure. Therefore a short-term exposure to a high-level concentration may result in acute effects (acute poisoning), whereas exposure to a low concentration spread over a long period of time, which would result in the same absorbed amount of the toxic substance, may be tolerated but may result in an even higher cumulative dose resulting in chronic effects.

2.2.3. Combined effects of chemicals

Occupational exposure is rarely confined to a single chemical. In most instances workers are exposed to two or more chemicals. The combined effect of multiple exposure to chemicals is an area for which sufficient information is often lacking. It may happen that a combination of two chemicals,

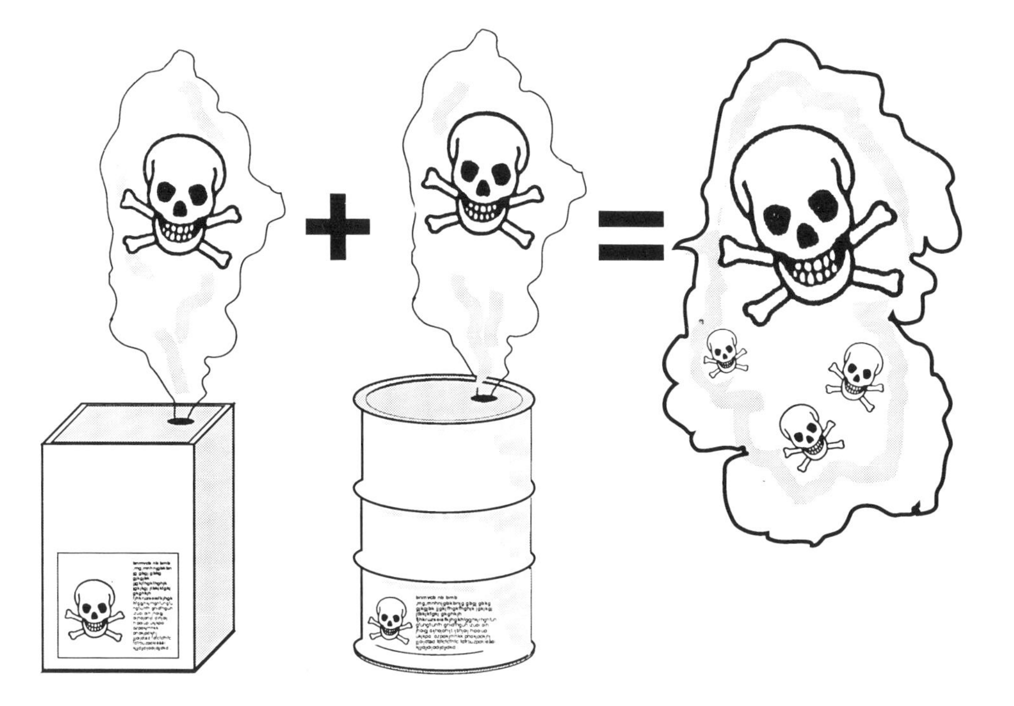

Figure 6.
The toxic effect of a combination of two chemicals may be far greater than the sum of the toxic effects of each

by chemical reaction or absorption into the body, produces a new substance with totally different properties and even more harmful to health than the chemicals acting separately (figure 6).

Because of this lack of information on the combined effects of chemicals, multiple exposure should be avoided or reduced to the lowest possible level.

> **Remember:**
> *Avoid mixing several chemicals together. The combination may result in very dangerous effects.*

2.2.4. Hypersusceptible groups

There is great variation in individual response to a chemical. Exposure to a particular dose over a similar time period will induce different responses among different people. Some may be severely affected and some may be mildly affected, while others may show no apparent effects. Individual sensitivity may also depend on age, sex and general state of health. Children, for example, will be more sensitive than adults. The unborn foetus may be very susceptible to the risks of chemical substances. Therefore, in the recognition of potential hazards, individual variation in sensitivity should be taken into account.

Discussion questions:

1. Are any workers in your workplace exposed to several chemicals at the same time?
2. List the chemicals and the forms in which they are used.
3. Are there groups of workers who might be especially susceptible to the effects of chemicals in your workplace?

2.3. The toxic effects of chemicals

As explained above, the effects of chemicals can be either acute or chronic, depending on the concentration and length of exposure. Chemicals may also produce different effects for different modes and types of exposure. The effects of chemicals can be categorized into the following groups:

- causing irritation;
- allergies;
- lack of oxygen;
- systemic poisoning;
- cancer;
- damage to the unborn foetus;
- effects on future generations;
- pneumoconiosis (dusty lung).

2.3.1. Irritation

Irritation means a condition that is aggravated when chemicals come into contact with the body. The parts of the body normally affected are the skin, the eyes and the respiratory tract.

2.3.1.1. Skin

When certain chemicals come into contact with the skin, they can rèmove the protective layer causing it to become dry, rough and sore. This condition is called dermatitis (figure 7). There are many chemicals causing dermatitis.

Figure 7. Many chemicals cause dermatitis by coming into contact with the skin

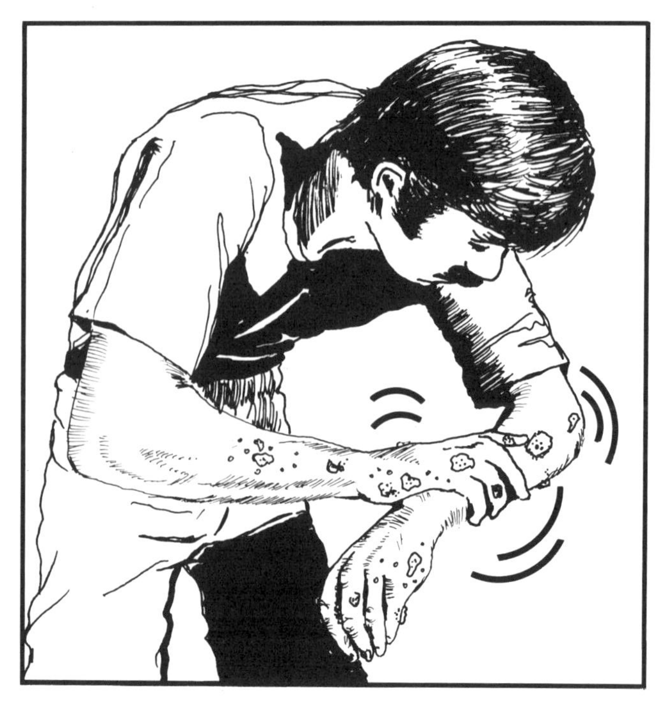

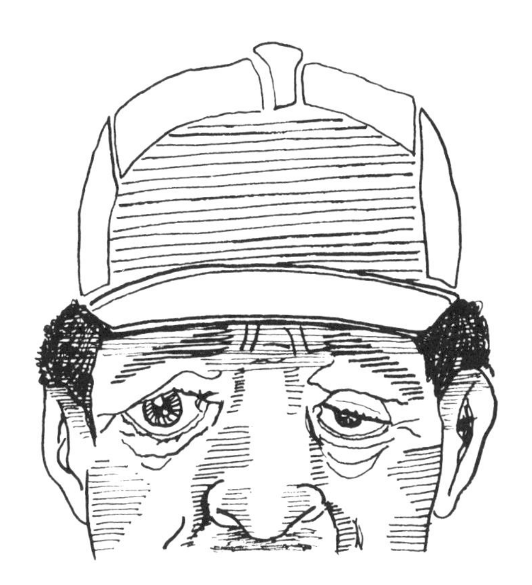

Figure 8. Toxic chemicals can irritate the eyes

2.3.1.2. Eyes

Contact of chemicals with the eyes may cause skin damage ranging from mild temporary discomfort to permanent damage (figure 8). The severity of damage depends on the dose and the speed with which first-aid measures are administered. Examples of substances causing eye irritation are acids, alkalis and solvents.

2.3.1.3. Respiratory tract

Irritants in the form of mist, gas or vapour will induce a burning sensation when in contact with the *upper respiratory tract* (nose and throat). This is normally caused by soluble substances such as ammonia, formaldehyde, sulphur oxide, acids and alkalis which are absorbed by the moist lining of the nose and throat. Care should be taken that these vapours are not inhaled when workers are dealing with chemicals, for example during spraying (figure 9).

Some irritants exert their effects along the *conducting airways*, causing bronchitis and sometimes serious damage to the lining and the lung tissues. Examples are sulphur dioxide, chlorine and coal dust.

Chemicals which are less soluble in water will penetrate to the *gas exchange areas*, causing serious irritation effects. Their presence in the workplace is not normally detected and can present serious hazards to workers. The reaction of chemicals with the lung tissues induces pulmonary oedema (fluid in the lungs), either immediately or after a few hours. The symptoms begin with intense irritation followed by coughing, dyspnoea (shortness of breath), cyanosis (lack of oxygen) and expectoration of large quantities of mucus. Examples are nitrogen dioxide, ozone and phosgene.

Discussion questions:

1. Describe how to recognize an irritant chemical in your workplace.
2. Are irritants properly labelled?

2.3.2. Allergy

An allergy can be acquired through exposure to chemicals. Initially workers may not develop an allergy; constant exposure, however, may produce a reaction by the body. Even exposure to a low level of a substance could later induce an allergic reaction. Either the skin or the respiratory tract can be affected by an allergic reaction.

2.3.2.1 Skin

An allergic reaction of the skin is a condition that often looks like dermatitis (small pimples or

Figure 9. When spraying, care should be taken that toxic vapours are not inhaled

watery blisters). This effect may not appear at the site of contact but can occur elsewhere on the body. Examples of sensitizers are epoxy resin, amine hardeners, azo dyes, coal tar derivatives and chromic acids.

2.3.2.2. Respiratory tract

Sensitization of the respiratory tract causes occupational asthma. The symptoms of this reaction often include coughing, especially at night, and difficulties in breathing such as wheezing and shortness of breath. Examples of chemicals causing this type of reaction are toluene diisocynate and formaldehyde.

> **Remember:**
> *Repeated exposures to chemicals may lead to allergic reactions.*

2.3.3. Lack of oxygen (asphyxiation)

Asphyxiation refers to interference with the oxygenation of the body tissues. There are two types: simple and chemical asphyxiation.

2.3.3.1. Simple asphyxiation

This refers to a condition whereby oxygen in the air is replaced by an inert gas such as nitrogen, carbon dioxide, ethane, hydrogen or helium to a level where it cannot sustain life. Normal air contains about 21 per cent of oxygen. If this concentration falls below about 17 per cent, the body tissues will be deprived of their supply of oxygen, causing symptoms such as dizziness, nausea and loss of coordination. This type of situation may occur in confined workplaces (figure 10). A further reduction of oxygen may cause unconsciousness and death.

2.3.3.2. Chemical asphyxiation

In this situation direct chemical action interferes with the body's ability to transport and use oxygen. An example of a chemical asphyxiant is carbon monoxide. Concentrations of 0.05 per cent of carbon monoxide in the air may considerably reduce the capacity of the blood to carry oxygen to the various tissues of the body. Another example is the toxic effect of hydrogen cyanide or hydrogen sulphide. These substances interfere with the cells' ability to accept oxygen even though the blood is rich in oxygen.

Discussion questions:

1. Describe areas in your workplace that may have an inadequate supply of oxygen.
2. List the special measures taken to inform, train and give special consideration to those working in low-oxygen areas.

2.3.4. Narcosis and anaesthesia

Exposure to relatively high concentrations of certain chemicals such as ethyl and propyl alcohols (aliphatic alcohol), acetone and methyl-ethyl ketones (aliphatic ketone), acetylene hydrocarbons, and ethyl and isopropyl ethers depresses the central nervous system. These chemicals will induce an effect similar to being drunk. Single exposure to a high concentration may result in unconsciousness or even death. There are also cases where workers have become addicted to these substances.

Figure 10. In certain working environments such as confined spaces, an inadequate amount of oxygen can lead to simple asphyxiation and death

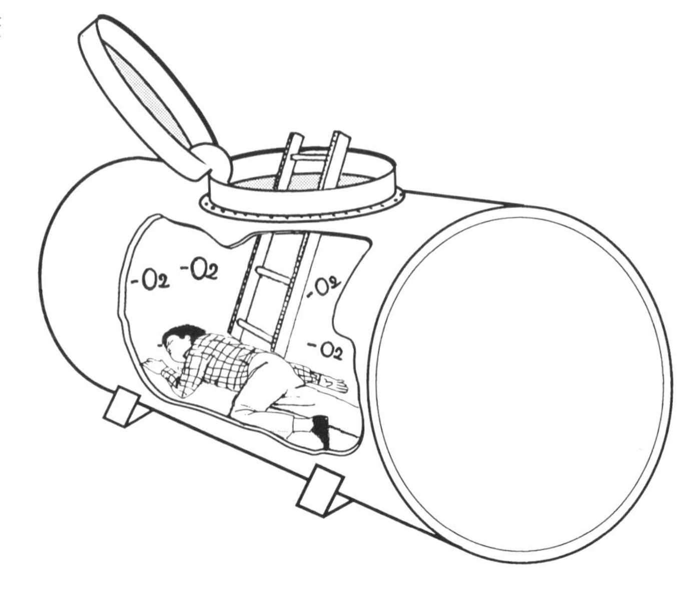

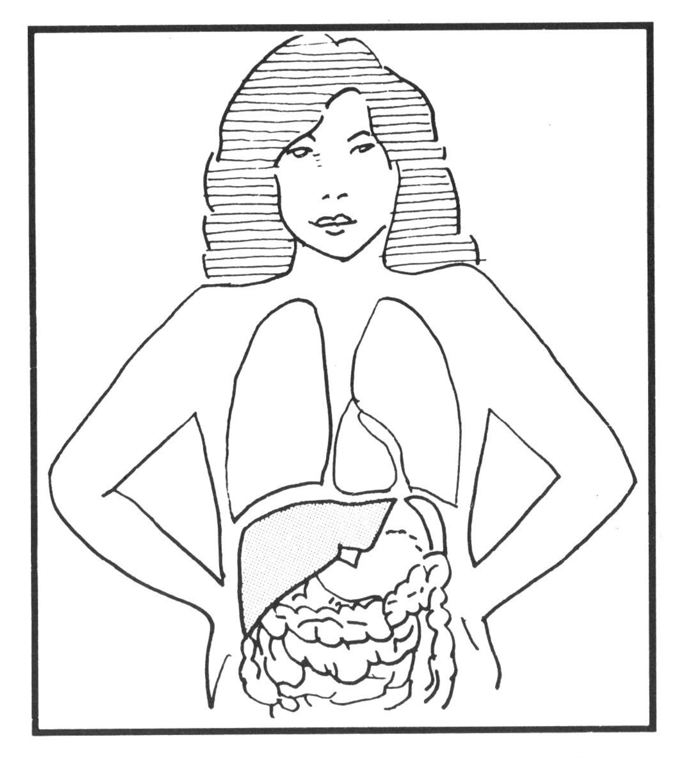

Figure 11. The liver is a target organ that may be damaged by exposure to certain chemicals

2.3.5. Systemic poisoning

The human body is made up of many systems. Systemic poisoning refers to the adverse response induced by chemicals to one or more body systems, which in turn spreads throughout the body. The effect is not localized at any one spot or area of the body.

One of the tasks of the *liver* is to purify any noxious substances from the blood by converting them to harmless and water-soluble substances before being excreted (figure 11). However, some chemical substances cause damage to the liver. Depending on the dose and frequency of exposure, repeated damage to liver tissues may cause injury resulting in scarring (cirrhosis) and decreased liver function. Liver injury can be caused by chemicals such as solvents (alcohol, carbon tetrachloride, trichloroethylene, chloroform) and may be mistaken for hepatitis, as the symptoms (yellowish skin and eyes) produced by these chemicals are similar.

The *kidneys* are part of the urinary system. Their task is to excrete waste products generated by the body, maintain the balance of water and salts, and control and maintain the acidity level of the blood (figure 12). Chemicals that prevent the kidneys from excreting poisonous products include carbon tetrachloride, ethylene glycol and carbon disulphide. Other chemicals such as cadmium, lead, turpentine, methanol, toluene and xylene will slowly deteriorate the kidney function.

The *nervous system* controls body function (figure 13), and can be damaged by certain chemicals. Chronic exposure to solvents has been linked to symptoms such as fatigue, sleep difficulties, headache and nausea. More serious cases cause motor disturbances, paralysis and an impaired sense of perception. Exposure to hexane, manganese and lead has been associated with effects on the peripheral nerves resulting in symptoms of "wrist drop". Exposure to organophosphate compounds such as parathion may cause the nervous system to fail. Another example is carbon disulphide, which has been linked with cases of mental disorder (psychosis).

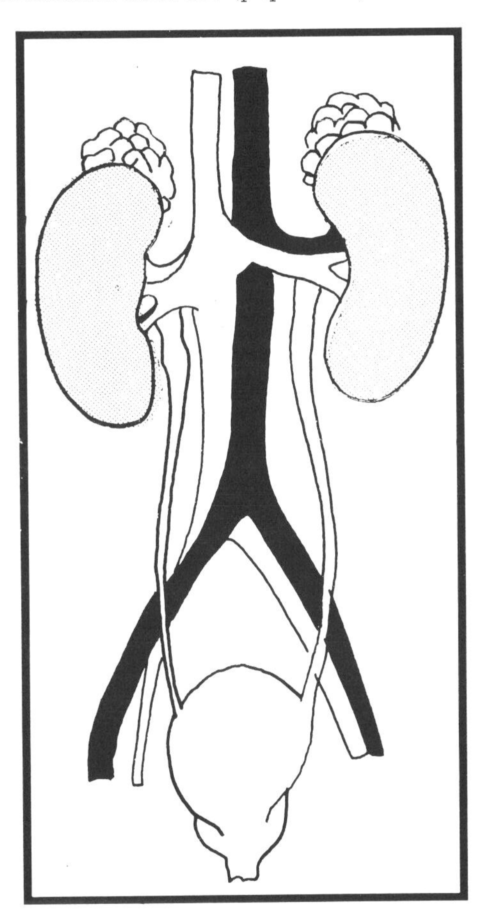

Figure 12. Certain chemicals may impede the normal functions of the kidneys

Exposure to certain chemicals may also have negative effects on the *reproductive system*, producing sterility in men and causing miscarriages in pregnant women. Chemicals such as ethylene dibromide, benzene, anaesthetic gases, chloroprene, lead, organic solvents and carbon disulphide have been linked with the reduction of fertility in male workers. Miscarriages are linked with exposure to anaesthetic gas, mercury ethylene oxide, glutaraldehyde, chloroprene, lead, organic solvents, carbon disulphide and vinyl chloride.

2.3.6. Cancer

Long exposure to certain chemicals may cause an uncontrolled growth of cells, resulting in cancerous tumours. These tumours may appear many years after the first exposure to the substances. This period is called the latency period and may range from four to 40 years. The site of occupational cancer varies and may not necessarily be confined to the contact area. Substances such as arsenic, asbestos, chromium, nickel and bischloromethyl ether (BCME) may cause lung cancer. Cancer of the nasal cavities and sinuses is caused by chromium, isopropyl oils, nickel, wood and leather dust. Cancer of the bladder is linked with exposure to benzidine, 2-naphthylamine and leather dust. Skin cancer is linked to exposure to arsenic, coal tar and petroleum products. Liver cancer can be caused by exposure by vinyl chloride monomer, while cancer of the bone marrow is caused by benzene.

Figure 13. The nervous system, composed of the brain, the spinal cord and the nerves, controls body functions and can be affected by chemicals

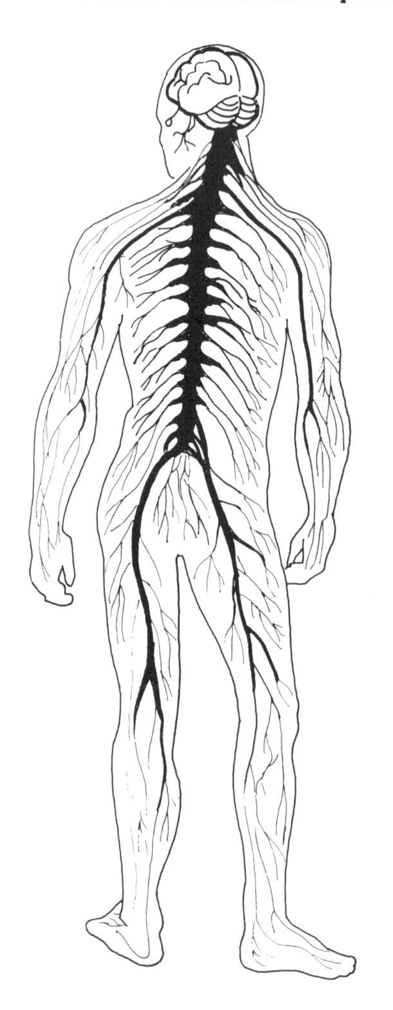

2.3.7. Damage to the unborn foetus (teratogenesis)

Congenital malformation resulting from exposure to chemicals may interfere with the development of the normal foetus. During the first three months of pregnancy, the vital organs such as brain, heart, arms and legs are forming. Studies conducted have suggested that in the presence of certain chemicals such as anaesthetic gases, mercury and organic solvent, the normal process of cell division may be interfered with, causing deformity in the foetus.

2.3.8. Genetic effects on future generations (mutagenesis)

The genetic effects of certain chemicals on workers may lead to undesired changes in future generations. Information on these effects is scarce. However, results of laboratory tests suggest that 80 to 85 per cent of carcinogenic chemicals may also have effects on future generations.

2.3.9. Dusty lung (pneumoconiosis)

Dusty lung, or pneumoconiosis, is a condition caused by the deposit of small dust particles in the gas exchange areas of the lung and the reaction of the tissue to their presence. Changes in the lungs are extremely difficult to detect at the early stage, and deterioration occurs long before such changes can be detected by X-rays. With pneumoconiosis the capability of the lungs to absorb oxygen will be reduced and the victim will develop shortness of breath during strenuous activities. The effect is irreversible. Examples of substances causing pneumoconiosis are crystalline silica, asbestos, talc, coal and beryllium.

Discussion questions:

1. Can you describe any situations in your workplace where a worker has become gravely ill as a result of an exposure to toxic chemicals?
2. With the help of chemical safety data sheets, locate and discuss the toxic effects of at least four commonly used chemicals in your workplace. What systems of the body can these chemicals affect? Are any target organs mentioned?
3. Describe special measures that can be observed to prevent exposure to these toxic chemicals.

Suggested further reading

ILO. *Encyclopaedia of occupational safety and health*, 2 vols. (Geneva, 3rd ed., 1983).

—. *Safety and health in the use of agrochemicals: A guide* (Geneva, 1991).

Patty's industrial hygiene and toxicology, 4 vols. (New York, Wiley-Interscience, 3rd ed., 1981).

3. Fire and explosion hazards

Chemicals in the workplace may pose certain risks of fire and explosion. The improper storage, transport, use or disposal of these chemicals may result in an event ranging from a minor fire to a disaster causing major economic loss, suffering and loss of human life.

This chapter will address some of the risks of fire and explosion due to chemicals in the workplace. Chapter 4 discusses some preventive actions. Above all, it is important to remember that the prerequisite to understanding the safe use of chemicals in the workplace is an adequate knowledge of them through chemical safety data sheets and easy identification through proper labelling.

This chapter will also help you understand some of the basic fire and explosion information that is found on the chemical safety data sheet.

In order for human beings to survive, at least three basic elements are essential: food, oxygen and heat. These elements must also be in the right proportions for human survival. Too much or too little food, oxygen or heat may result in discomfort, illness or death.

In the same way, fire needs three elements: fuel, oxygen and a source of heat. These must be in the right proportions and the right states before fire can be ignited and before it can continue to burn. The fuel must be at a temperature, defined as the flashpoint, at which flammable vapours are released. Therefore there must be enough heat to bring the fuel to this point. There is also a need for oxygen. Normally, fire needs 15-21 per cent oxygen to ignite and burn.

Discussion questions:

1. Have there been any fires or similar dangerous situations involving chemicals in your place of work?
2. What happened and what were the results?

3.1. Fuel

When examining the risks of fire and explosion resulting from hazardous chemicals, one must first look at the chemical and its characteristics. Most often the chemical substance will act as the source of fuel in the fire or explosion triangle (figure 14).

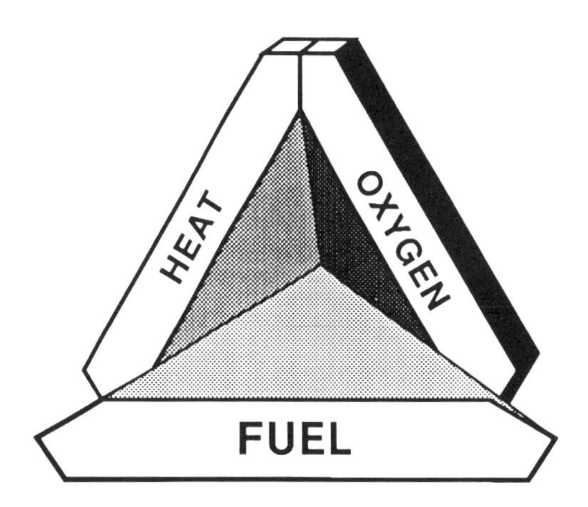

Figure 14. Fuel is the first element of the fire and explosion triangle

3.1.1. Liquid flashpoints

One characteristic of chemicals that pose a fire or explosion risk is the flashpoint. This is the lowest temperature at which a chemical gives off flammable vapours. It is the flammable vapour that burns, not the liquid. Table 1 shows the flashpoints of some commonly used chemicals.

Table 1. Some common flashpoints

Chemical	Flashpoint (° C)
Gasoline	−43
Acetone	−19
Methyl alcohol	11
Kerosene	43
Heptanc	−4
Toluene	6

> **Remember:**
> *In terms of fire and explosion, the lower the flashpoint the more dangerous the chemical.*

Other factors may relate to a chemical's ability to reach its flashpoint. For example, when a liquid such as kerosene is atomized, it will produce flammable vapours which will burn at a lower ambient temperature than its flashpoint. In addition, a chemical with a high flashpoint may

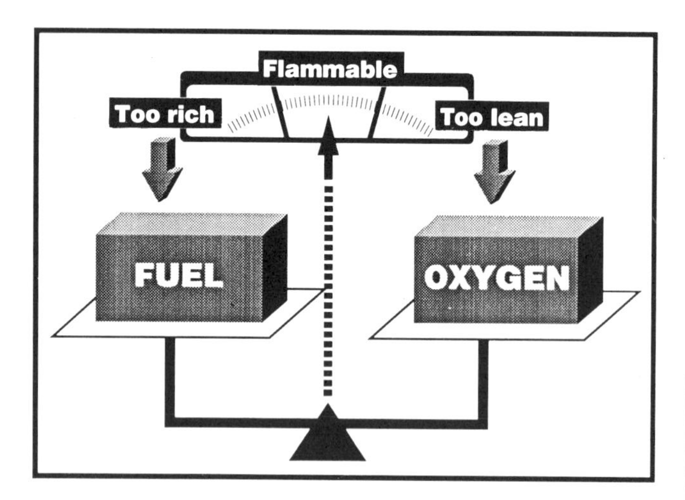

Figure 15. A flammable liquid will start burning only if the fuel and oxygen are in the right proportions

be heated to its flashpoint by other substances with lower flashpoints burning in close proximity to the first substance. It is therefore essential that careful consideration should be given to the safe storage of hazardous chemicals.

When a chemical is heated to its flashpoint, the flammable vapours will burn only until the heat source is removed. However, once the liquid reaches its firepoint (normally only a few degrees above the flashpoint), the vapours will continue to be produced and burn. The flashpoint is normally found on the chemical safety data sheet.

3.1.1.2. Upper and lower explosive limits

Flammable liquids must also have the right mixture of oxygen to burn (figure 15). An excess of fuel with an inadequate amount of oxygen may mean that the substance is too rich to burn. Conversely, a high concentration of oxygen with an inadequate amount of fuel may mean that the substance is too lean to burn. The limits at which a substance will ignite, depending on the percentage of oxygen, are called the upper and lower explosion limits (UEL and LEL). The UEL and LEL are normally found on the chemical safety data sheet.

3.1.1.3. Vapour weight

An additional consideration is the weight of the chemical as compared with the relative weight of air. This is known as the vapour weight. Gasoline vapours, for example, are three-and-a-half times heavier than air. Other chemical substances whose vapour weights are heavier than air include kerosene, carbon disulphide, acetylene and carbon monoxide. When working with a chemical that is heavier than air, it is important to know that its vapours may travel long distances and form concentrations at a considerable distance from their source, often at a low point such as a cellar.

Remember:
Vapours of chemicals that are heavier than air may travel long distances and concentrate in cellars.

3.1.2. Solids

Certain chemicals, in a solid state, will burn rapidly once ignited. Magnesium, for example, will burn once ignited and will be very difficult to extinguish.

Fuels in the form of dusts or powders may also explode in the right mixture of oxygen. When agitated and an appropriate ignition source is present, these dusts and powders will burn explosively creating multiple sequential explosions as additional dusts or powders are agitated.

3.1.3. Gases

Gases, of which several are very flammable, are commonly used in industry. Acetylene, hydrogen and methane (often a by-product) will explode in the right concentration of gas and oxygen when an ignition source is present.

Caution must also be exercised in dealing with compressed gases stored in pressurized containers. These gases, when heated within the containers, may expand to a point where the container fails, frequently resulting in a disastrous situation.

Discussion questions:

1. Using a chemical safety data sheet, find the flashpoints, the upper and lower explosive limits, and the vapour weight for four

chemicals commonly used in your workplace. Do you consider any of these chemicals particularly hazardous?

2. List and discuss the hazardous properties of chemical gases that are used or found in your workplace.

3.2. Heat

Heat is the second element of the fire or explosion triangle (figure 16). It is needed to bring the fuel to its flashpoint (if the flashpoint is above the ambient temperature) and to ignite the flammable vapours. Sources of heat that can ignite hazardous chemicals include electrical current, static electricity, spontaneous combustion, chemical reaction, friction, process heat, open flames, solar heat, radiant heat (hot surfaces) and lightning. One of the keys to the prevention of fires and explosions caused by hazardous chemicals is the control of sources of heat. The control aspects will be discussed in Chapter 4.

3.2.1. Electrical current

Heat is generated through electrical current in three ways: resistance, arcing and sparking. *Resistance* occurs when electricity travels through wires not large enough to carry the current. The result is either a blown fuse or a tripped circuit breaker, or the heating of the wire in the circuit. This circuit may reach a high enough temperature to ignite flammable vapours present in the air or may cause combustible materials to ignite, burn and elevate the temperature of nearby chemicals to their flashpoint and their firepoint. Electrical *arcing* happens when electrical current jumps from one point to another. This may occur in a switch or connection box when wires separate from connectors or when the insulation of wires is worn away between a positive and a neutral wire (e.g. when temporary wiring or extension cords have been exposed to forklift trucks or when workers constantly tread on the wire and wear away the insulation). The resulting electrical arc can ignite flammable vapours. Molten metal resulting from the arcing can also ignite combustible materials, thus causing the heating of chemicals as above. *Sparking* may also ignite flammable vapours that are present.

3.2.2. Static electricity

Static electricity is generated when two dissimilar surfaces come together and are separated, resulting in the build-up of positive and negative charges (figure 17). The resulting spark may cause the ignition of flammable vapours or an explosion. For example, in machines which process film and sheet material, insulating

Figure 16. Heat is the second element of the fire and explosion triangle

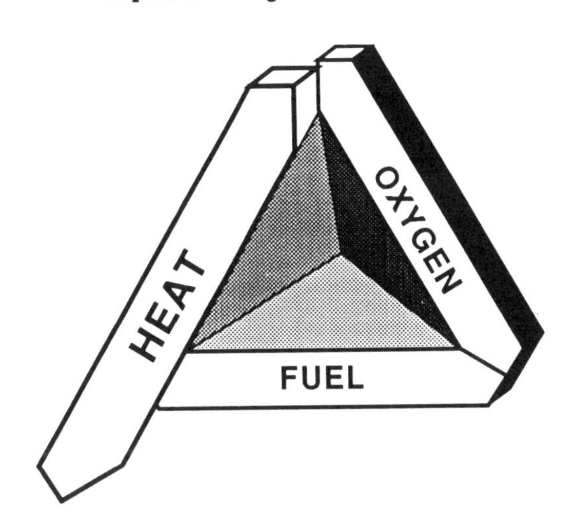

Figure 17. When two dissimilar surfaces come together and are separated, the result may be an electric charge

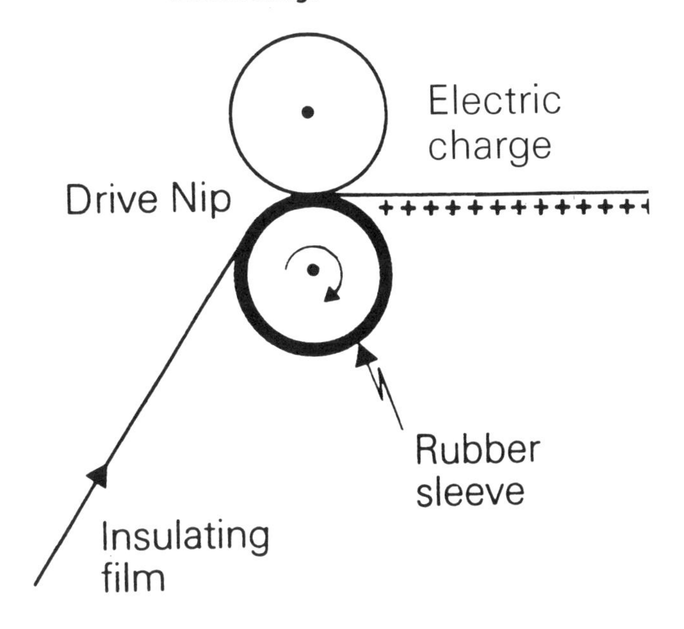

material becomes charged by passage through a machine (figure 17). If such materials continue to be processed in flammable atmospheres, the charges created should be carefully neutralized to avoid sparking. Static build-up can occur when two surfaces rub together, or when liquids are transferred from one container to another without proper earthing and bonding (a common source of explosions when flammable liquids are transferred).

3.2.3. Spontaneous combustion

This phenomenon has been known to occur in industry when piles of oily rags have been left to dry in the open air. Certain kinds of oils tend to produce heat, as they are oxidized, and may create a fire in the pile of rags. (A similar situation may occur in agriculture by the heat created by fermentation when wet hay is baled and stored.) The simple measure of storing oily rags in covered containers (thus reducing the amount of oxygen) diminishes this risk.

Figure 18. Mixing two or more chemicals together may produce heat

3.2.4. A mixture of two chemicals

As stated in Chapter 2, when two or more chemicals are mixed, the combined effect can be more dangerous than the sum of their separate effects. This combined effect may also carry a greater risk of fire and explosion. For example, the mixed chemicals may have a lower flashpoint and a lower boiling point, and may give off easily ignitible vapours.

The reactions of two chemicals coming together could possibly produce sufficient heat, as a by-product, for other chemicals in the vicinity to be heated to a point where they become dangerous (figure 18). Thus a chain reaction could be started that could have catastrophic results.

3.2.5. Friction

When two surfaces rub together, heat may be produced. This is known as friction. Drive belts rubbing against their housings or guards, or metal surfaces rubbing against each other, may generate an adequate amount of heat to ignite flammable vapours. Friction is frequently due to a lack of adequate maintenance, resulting in loose guards or inadequately lubricated metal surfaces or joints. A spark can also occur when a stone lodged in the sole of a shoe strikes a concrete surface.

3.2.6. Radiant heat

Heat from furnaces, vats, cooking stoves and other hot surfaces may ignite flammable vapours. The normal manufacturing process may also cause the production of sufficient heat to bring the chemicals stored in the vicinity to their flashpoint and to ignite the vapours. The direct rays of the sun, either by themselves or magnified by plastic or glass, can also have this effect.

Remember:

Production of heat may cause chemicals to reach a temperature where flammable vapours are present. The heat may also ignite the flammable vapours, causing a fire or explosion.

Figure 19. The open flame from welding and cutting torches can cause the ignition of flammable vapours

3.2.7. Open flames

Unprotected flames caused by cigarettes, matches, welding torches and internal combustion engines are very important sources of heat. When coupled with adequate fuel and in the presence of oxygen, they can create a fire or explosion (figure 19).

Discussion question:

List at least three sources of heat that may ignite flammable liquids in your workplace.

3.3. Oxygen

Oxygen is the third element of the fire or explosion triangle (figure 20). Most fuels need at least 15 per cent oxygen to burn. In excess of 21 per cent, oxygen may cause more rapid intense combustion leading to explosions. Sources of oxygen other than the environment may include oxygen cylinders for cutting and welding operations, oxygen supplied by piped-in manifolds for process operations, and at times chemical reactions. Oxygen released by a chemical when heated is known as an oxydizer; some examples are given in table 2.

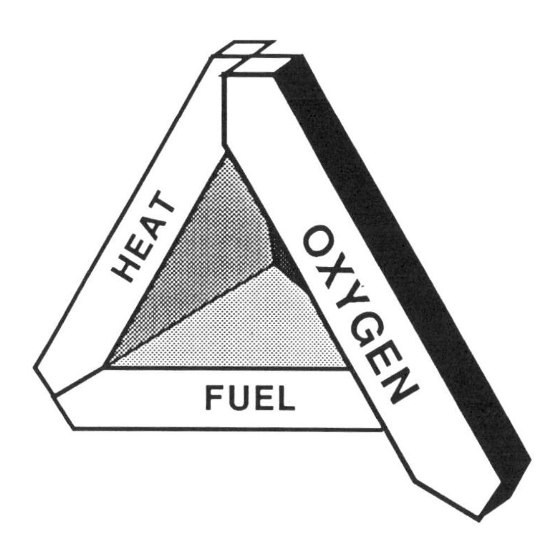

Figure 20. Oxygen is the third element of the fire and explosion triangle

Table 2. Examples of chemicals that give off oxygen when heated

Nitrates	Ammonium nitrate and sodium nitrate
Nitrites	Ammonium nitrite
Inorganic peroxide	Hydrogen peroxide
Permanganates	Potassium permanganates

Discussion questions:

1. Are there any common sources of oxygen in your workplace that can intensify the burning of a flammable liquid?
2. How can these sources of oxygen be controlled?

Suggested further reading

National Fire Protection Association: *Fire prevention handbook* (Quincy, Massachusetts, 17th ed., 1991).

4. Basic principles of prevention

4.1. Four principles of operational control

The general objective in the control of hazards relating to chemicals in the workplace is to eliminate or reduce to the lowest possible level the hazardous chemicals that can come into contact with the worker or the environment, or that can produce a fire or an explosion.

To achieve this, a four-point strategy of operational control is used to prevent or reduce the possibility of exposure to chemicals, and thus decrease the risk of accidents, and fires and explosions due to chemicals.

Ideally, the best method of preventing diseases and injuries, and fires and explosions caused by chemicals would be to provide a working environment free from such chemicals. However, this is not always feasible. Therefore, it is necessary to isolate the danger, increase ventilation or use personal protective equipment. But first it is essential to identify the chemical hazard and the extent of its risk, and to examine the inventory, storage, process of transfer and handling, and the actual use of the chemical and its disposal. In dealing with each specific hazard, the following four points should be considered as a prevention strategy:

Four principles of operational control

1. *Eliminate the hazard:* Eliminate the dangerous substance or process or replace it by a less dangerous one.
2. *Put distance or shielding between the substance and the worker:* Prevent the dangers associated with the chemical from reaching the worker.
3. *Ventilation:* Provide general and local ventilation to remove or reduce the concentration of airborne contaminants such as fumes, gases, vapours and mists.
4. *Protect the worker:* Provide personal protection to the worker to prevent physical contact with the chemical.

These principles are discussed in detail below.

4.1.1. Elimination or substitution

The most efficient way to reduce chemical hazards is to avoid using toxic substances or substances that pose a risk of fire or explosion. The selection of the chemical substance should be made at the design and planning stage of the process. For existing processes, the substitution method should be used wherever hazardous

Figure 21.
Wherever possible, hazardous substances should be replaced by less hazardous ones. For example, an organic solvent-based glue should be replaced by a water-based one

substances or processes may be replaced with others that are less hazardous.

Examples of substituting for toxic substances are: using water-based paint or glues instead of those with an organic solvent base (figure 21); using water-detergent solutions instead of solvents; using trichloromethane as a degreasing agent instead of trichloroethylene; and using chemicals with higher rather than lower flashpoints. Examples of process substitution are: replacing spray painting with electrostatic or dip painting; replacing manual batch charging with continuous mechanical hopper charging; and replacing dry abrasive blasting with wet blasting.

Remember:
Try to reduce the risk by eliminating the chemical hazard or replacing the chemical with a less hazardous one.

The choice of alternative substances may be limited, especially where the use of that particular substance is unavoidable if the technical and economic requirements are to be met. It is always useful to look for alternative substances by learning from experiences in similar circumstances.

Discussion questions:

1. Are there chemicals in your enterprise which you think could be replaced by less dangerous ones?
2. Which organizations or agencies do you think can help you obtain information about possible substitutes for hazardous chemicals?

4.1.2. Putting distance or shielding between the substance and the worker

This method involves enclosing processing equipment in order to restrict the spread of air contaminants to the workplace environment and isolating sources of heat from open flames or fuels. It is ideal for processes in which the worker has minimal chances of coming into contact with the chemical in question. Examples of this method are screening the whole machine, enclosing dust-producing transfer points of conveyors, or shielding abrasive blasting operation processes.

Contact with hazardous chemicals can also be reduced by isolation, which entails moving the hazardous processes or operations to a remote location in the plant or constructing a barrier to separate them from other processes (figure 22).

Examples of isolation are abrasive blasting of huge structures in a remote area of the workplace, or separating a spray painting process from the other processes of the plant by means of a barrier or wall.

A similar isolating effect can be obtained by the safe storage of hazardous chemicals and by

Figure 22. Using remote control may prevent the dangers associated with the chemical from reaching the worker

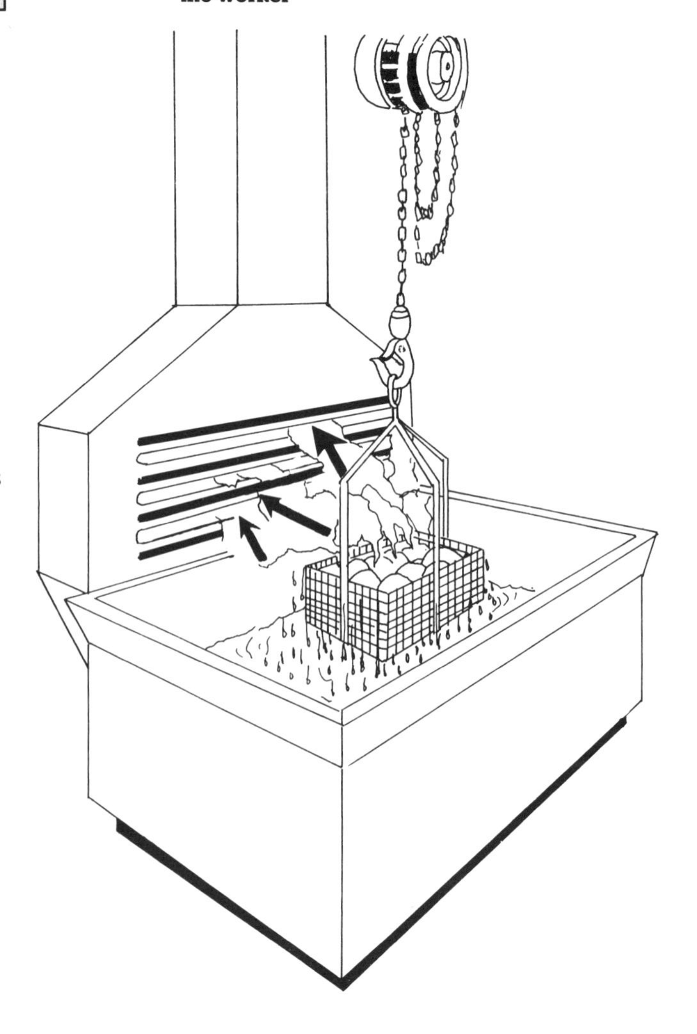

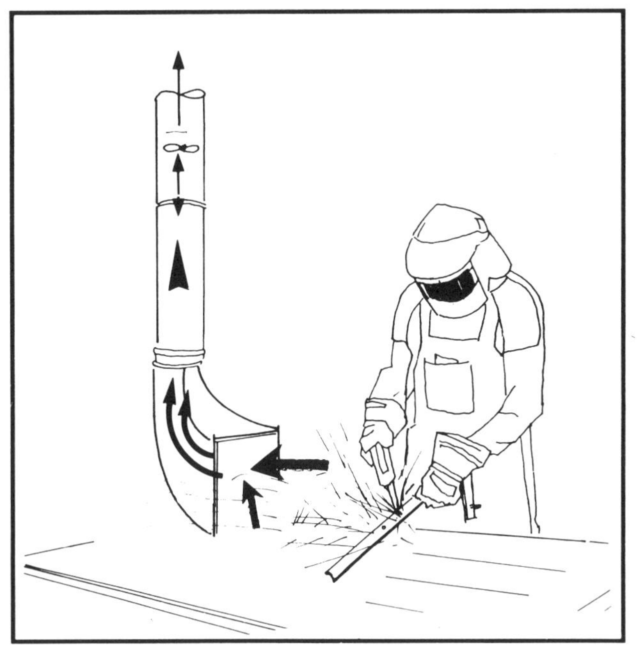

Figure 23. Here are two methods of local ventilation. On the left, the contaminants are drawn into the work table before they can reach the breathing zone of the worker. On the right, fumes from welding are drawn into an exhaust system

restricting the amounts of these chemicals in the workplace to that required in a day or a shift. Such restriction is useful if the process can be carried out by a very small number of workers and when control by other methods is difficult or impossible. The workers engaged in this process, however, should be given adequate personal protection.

4.1.3. Ventilation

In the case of airborne chemicals, ventilation is regarded as one of the best forms of control apart from substitution and enclosure. By means of adequate ventilation, we can trap contaminants released into the air from the process or operation and prevent them from entering the breathing zone of the worker. The trapped contaminants are conveyed by ducts to a collector (cyclone, filter house, scrubbers or electrostatic precipitators) where they are removed before the air is discharged into the outside environment. This is accomplished by a special exhaust system or by increasing the general ventilation.

Figure 23 shows two methods of local ventilation. In the case of an exhaust ventilation system, hoods at the point where the air enters the system must be located as close as possible to the source of the contaminant, otherwise the draught induced by the system's fan will not be strong enough to capture the contaminant. In order to ensure the efficiency of an exhaust ventilation system, it is important to check its design by getting advice from a specialist or person trained in installing ventilation systems. The system should be regularly maintained to keep it operating efficiently. Exhaust ventilation systems have been used effectively to control toxic substances such as lead, asbestos and organic solvents.

General ventilation is also known as dilution ventilation. It works by diluting the air contaminants or the concentration of flammable vapours through pushing air to and from the workplace. This system uses natural air movement from open windows or doors, or from a mechanical air-moving device. The airflow should be taken into account in the design of a building (figure 24). Because it tends to disperse the contaminants instead of removing them, this system is recommended only for substances with low toxicity, non-corrosive and used in small quantities.

Figure 24. The design of a building may increase airflow and dilute the concentration of substances with low toxicity

Discussion question:

What kinds of local ventilation are used in your workplace? Are they effective?

4.1.4. Personal protective equipment

If it is not possible to reduce chemical hazards to an acceptable level, workers must be protected by using personal protective equipment. This equipment forms a barrier between the toxic chemical and the route of entry and does nothing to reduce or eliminate the hazard. Failure of the equipment therefore means immediate exposure to the hazard. Thus, personal protective equipment should not be regarded as a primary means to control hazards but rather as a supplement to other types of control measure. With the risks of fire and explosion there are no sure ways to provide equipment to protect the workers.

4.1.4.1. Respirators

Respirators, which cover the mouth and nose of the worker, prevent the entry of chemicals into the body by inhalation. The use of respirators should be confined to situations:

- where interim control measures are necessary before engineering controls are installed;
- where engineering controls are not practicable;
- to supplement engineering controls in reducing exposure during maintenance and repair;
- during emergencies.

Respirators should be selected according to the following criteria:

- the identification of the contaminant or contaminants;
- the maximum possible concentration of contaminants in the workplace;
- the acceptability to the worker in terms of comfort;
- compatibility with the nature of the job and the elimination of risks to health;
- a proper fit to the face of the user in order to prevent leakage.

They can be divided into two types: air-purifying and air-supply respirators.

Air-purifying respirators clean the air by filtering or absorbing contaminants before they enter the respiratory system. The cleaning devices are made of filters for removing dust from the air (figure 25), or chemical cartridges or canisters for absorbing gases, fumes, vapours and mists (figure 26). These respirators come in the form of half-face (covering the mouth, nose and chin) or full-face masks (covering the face including the eyes). There are no filter-type or cartridge-type

Figure 25. A dust-mask respirator

respirators which can protect the worker from all hazardous chemicals.

The types of filter and of cartridge or canister differ according to types of dust or of gaseous contaminants. It is essential to obtain advice on the proper type of respirator for a given hazard from the supplier of the respirator.

Air-supply respirators provide a continuous supply of uncontaminated air and offer the highest level of respiratory protection. Air can be pumped either from a remote source (connected by a high-pressure hose) or from a portable supply (such as from a cylinder or a tank containing compressed or liquid air or oxygen). This portable type, illustrated in figure 27, is known as self-contained breathing apparatus (SCBA). The mask of air-supply breathing apparatus is designed to cover the whole face.

To ensure effective use, workers must be trained and educated in the proper use, care and maintenance of respirators (figure 28). Wearing

Figure 26. A half-face cartridge respirator

Figure 27. Self-contained breathing apparatus (SCBA)

Figure 28.

All workers who may need to use respirators should be regularly trained in their use, care and maintenance

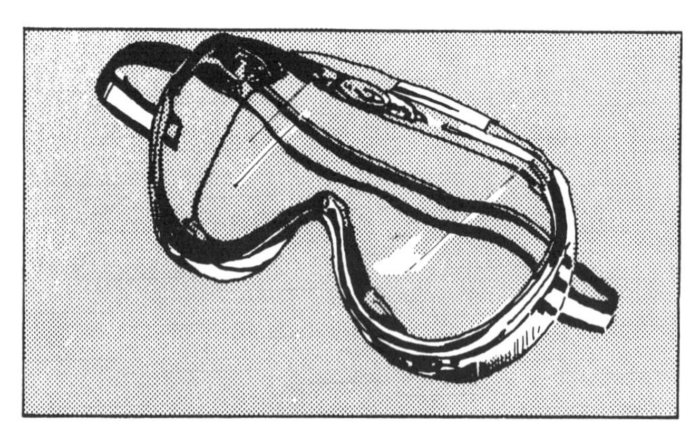

Figure 29. Protective goggles for eye protection

a poorly maintained respirator can be more dangerous than not wearing one at all. In this situation the workers think that they are protected while in reality they are not.

Figure 30. An eye and face shield

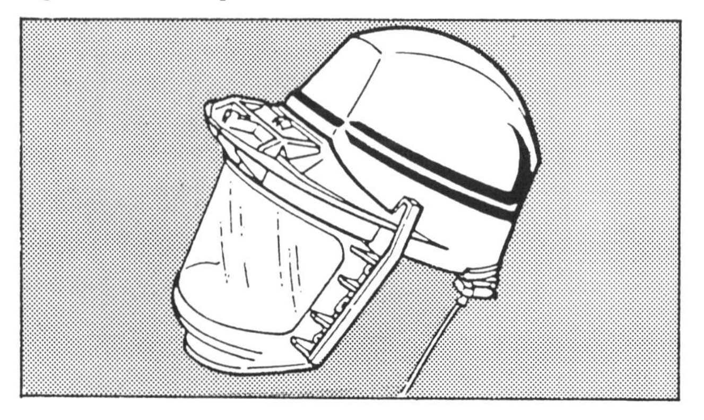

4.1.4.2. Other personal protective equipment

Protecting the eyes and skin from chemical splashes and exposure to dusts, vapours, mists and fumes may require appropriate protective equipment.

Examples of eye and face protection include safety glasses, safety goggles (figure 29) and face shields used to protect against corrosive liquids,

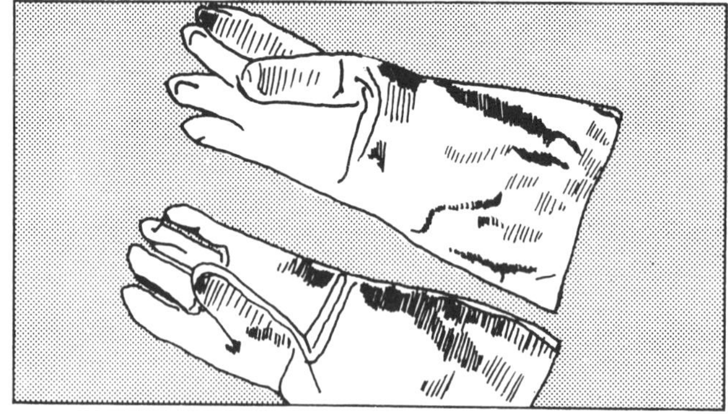

Figure 31. Gauntlet glove for hand protection

solids and vapours, and other foreign bodies (figure 30).

To protect the skin, protective clothing such as gloves, aprons, boots and coveralls, made from impervious materials, should be provided to eliminate prolonged or repeated contact with solvents or other chemicals. A wide range of materials is used to make these protective devices, and proper selection is essential. For example, cotton and leather gloves are suitable for protecting hands from dust, while rubber gloves are suitable for protection against corrosive substances (figure 31). The main consideration is that these materials offer resistance to chemicals. Suppliers of protective equipment should be able to give advice in this respect.

Protective creams and lotions are also available for protecting the skin. Their effectiveness varies, but if properly selected and applied they can be very useful. There is no all-purpose cream; some are made to protect against organic solvents while others are designed for water-soluble substances.

Discussion questions:

1. Under what circumstances in your workplace is personal protective equipment used?
2. Outline the training necessary for the safe selection, use and maintenance of respirators.

Remember:
Personal protective equipment should be appropriate to the hazard and great care should be taken to fit the equipment to the worker.

Figure 32. Thoroughly wash exposed parts of the body

Figure 33. Personal protective clothing should be washed after use

4.1.5. Personal hygiene

Personal hygiene aims to keep the body clean and not allow anything harmful to remain on it for long periods as it can be absorbed through the skin. It is equally important to avoid inhaling or ingesting small, even minute, quantities of chemicals because of their harmful effects on health.

The basic rules of personal hygiene in using chemicals at work are as follows:

- avoid exposure to chemicals by following safe practices and using protective clothing and equipment as described above;
- thoroughly wash exposed parts of the body after work, before eating, drinking or smoking, and after using the lavatory (figure 32);

- examine the body regularly to ensure that the skin is clean and healthy;
- provide a protective dressing to any part of the body where there are cuts or sores;
- avoid self-contamination at all times, particularly when decontaminating or removing protective clothing;
- do not carry contaminated items such as dirty rags or tools in the pockets of personal clothing;
- remove and wash separately any contaminated item of personal protective clothing daily (figure 33);
- keep fingernails clean and short;
- avoid working with any product which causes an allergic response such as a skin rash.

There are other hygienic measures to be observed:

- even if the product label does not recommend wearing protective clothing, remember to cover as much of the body as possible, e.g. a long-sleeved shirt;
- as protective clothing is uncomfortable to wear and work in, seek advice about chemicals that do not require the use of protective clothing. Read the label before purchasing and ask the supplier.

4.2. Organizational control

Organizational control refers to measures and procedures established by management as part of a programme to control exposure, or to monitor the effectiveness of other control measures. The following measures should be taken:

- identification of all hazardous chemicals used;
- labelling;
- provision and use of chemical safety data sheets;
- safe storage;
- procedures for safe transfer;
- safe practices for handling and use;
- housekeeping measures;
- disposal routines;
- monitoring of exposure;
- medical surveillance;
- record-keeping;
- training and education.

Further details of these aspects as part of a chemical control programme are given in Chapter 6.

4.2.1. Identification

The principles of hazard identification are to know what chemicals are used or produced, how these chemicals come into contact with the body and cause injury or disease, how they may cause a fire or explosion in the workplace or how a spill or leak may cause harm to the environment.

Every chemical substance in the workplace should be known and accompanied by an appropriate label and an up-to-date chemical safety data sheet. To obtain this information, the employer should first request it from the supplier of the chemical. If the supplier does not have the information, the employer should then seek advice from the government, laboratories, universities or other specialized institutions.

Figure 34. Every chemical in the workplace should have a label and sufficient information to ensure its safe use

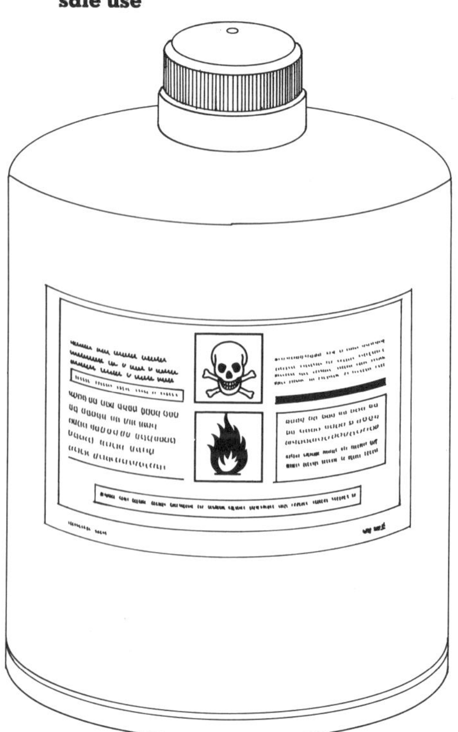

In fact, any chemical that cannot be identified, labelled or provided with a chemical safety data sheet should not be used until the relevant safety and health information has been obtained from the supplier and provided to workers and their representatives in a form and language which they can easily understand.

4.2.2. Labelling

Drums, bags and any other containers holding chemicals should be frequently checked for proper labels. The purpose of labelling is to warn workers of the potential hazards of the chemicals, the necessary precautionary measures and what to do in case of an emergency (figure 34). The following information should, as appropriate, be included on the label:

- trade names;
- identity of the chemical;
- name, address and telephone number of the supplier;
- hazard symbols;
- special risks associated with the use of the chemical;
- safety precautions;
- identification of the batch;
- the statement that a chemical safety data sheet giving additional information is available from the employer;
- the classification assigned under the system established by the competent authority.

When a hazardous chemical has been transferred from its original shipping container, the secondary and all subsequent containers should carry the appropriate warning labels. Labels should be affixed to all containers from the origins of the chemical to its neutralization or disposal. Any chemical that cannot be readily identified should be properly disposed of.

4.2.3. Chemical safety data sheets

Chemical safety data sheets should be available for every chemical substance within the enterprise. The data sheet provides basic information about the chemical and safety in its use (figure 35 and Annex 3). It also indicates appropriate precautions, including personal protective equipment, as well as emergency procedures.

Discussion questions:

1. How can a programme to assure proper labelling of every chemical in your workplace be carried out?
2. What actions should you take and whom should you notify if you find an unlabelled container containing a chemical substance?

> **Remember:**
> *Every chemical container in the workplace, no matter how small, should have an appropriate, understandable label.*

Figure 35. An example of a chemical safety data sheet, providing essential safety and health information

International Occupational Safety and Health Information Centre

Tel. + 41 22 799 67 40
Telex 415 647 ILO CH
Telefax + 41 22 798 62 53

International Labour Office

ILO-CIS CH-1211 GENEVA 22

CHEMICAL INFO-SHEET

CS-1 BENZENE

CAS 71-43-2
FORMULA: C_6-H_6

DESCRIPTION
Colourless liquid with sweet odour.
Used to produce:
- dyes
- plastics
- textiles
- detergents
- paints
- other chemicals

Used as a solvent for paints and adhesives.
Present in small amounts in gasoline.
Industrial uses are decreasing.

SHORT-TERM EXPOSURE EFFECTS
Very Toxic
Inhalation:
A 5-hour exposure at 50-150 ppm can cause:
- headache
- tiredness

A 1-hour exposure at 200-500 ppm can cause:
- nausea
- dizziness
- confusion

A 30-60 minute exposure at 3000 ppm can cause nose and throat irritation.
A 30-minute exposure at 7500 ppm can cause death.

Eye Contact:
High concentrations of vapour cause slight irritation.
Liquid causes a slight burning sensation.

Skin Contact:
Liquid dissolves skin oils and causes irritation and blistering.

Ingestion:
May cause the same symptoms as inhalation.
If swallowed, liquid drawn into lungs can cause severe injury.

LONG-TERM EXPOSURE EFFECTS
Benzene can damage the blood-forming system causing:
- anemia
- infections
- bruising
- bleeding

Prolonged low-level exposure can cause:
- hearing damage
- headache
- dizziness
- tiredness
- paleness
- problems with vision and balance

Repeated skin contact causes:
- redness
- drying
- blistering

Known to cause cancer in humans.
Cancers of the white-blood cells can develop.
Reproductive effects such as menstrual problems may result.
Genetic damage can develop after long-term, severe exposures.

FIRE AND EXPLOSION
Flammable Liquid
Highly flammable.
Dangerous fire hazard.
Extinguish fires with:
- dry chemical
- foam
- carbon dioxide

Vapours can travel at ground level to ignition source and flash back.

CHEMICAL REACTIVITY
Normally stable.
Contact with strong oxidizers, such as nitric acid, increases risk of fire and explosion.

PERSONAL PROTECTION
Inhalation:
Wear a self-contained breathing apparatus or a supplied-air respirator if vapour or mist concentration is unknown or present at any detectable concentration.

Skin:
Wear, as needed:
- gloves
- coveralls
- boots

A suitable material is Viton.
Have a safety shower/eye-wash fountain available in the immediate area.

Eyes:
Wear chemical safety goggles.
A face shield may also be necessary.

STORAGE AND HANDLING
Follow rules for storing and handling flammable liquids.
Store benzene:
- in tightly-closed, grounded, labelled containers
- in a cool, dry, well-ventilated area
- out of direct sunlight
- away from incompatible materials and heat.

Use non-sparking ventilation systems and electrical equipment.
Use in small quantities in designated areas.
Prevent release of vapours into workplace air.

CLEAN-UP AND DISPOSAL
Only trained personnel should clean up.
Ensure appropriate ventilation is provided.
Use appropriate protective clothing and respirators.
Stop or reduce leak if possible.
Absorb small spills with sand or other inert material.
Place in suitable, covered containers.
Flush area with water.
For large spills, contact emergency services and supplier for advice.
Comply with environmental regulations.

FIRST AID
Inhalation:
Remove source of benzene or move victim to fresh air.
If breathing has stopped, begin artificial respiration.

Eye Contact:
Flush affected eye with lukewarm, gently flowing water for 20 minutes, holding the eyelid open.
Do not rinse contaminated water into non-affected eye.

Skin Contact:
Remove contaminated clothing.
Gently blot or brush away excess chemical quickly.
Wash gently and thoroughly with water and non-abrasive soap.

Ingestion:
Never give anything by mouth if victim is:
- losing consciousness
- unconscious
- convulsing

Rinse mouth thoroughly with water.
Have victim drink about 250 mL (8 oz.) of water.
DO NOT INDUCE VOMITING.
If vomiting occurs, have victim lean forward and repeat administration of water.

***Note:** Obtain medical attention IMMEDIATELY for all serious exposures. Consult a physician or the nearest Poison Control Centre.*

NEED MORE INFORMATION?
See CHEMINFO record no. 179E, Chemical Hazard Summary No. 34, available from CCOHS.

This document was originally published by CCOHS (Canadian Centre for Occupational Health and Safety) in its **Chemical Infogram series**
Further information can be obtained from CIS or its national centres.

The following information is normally included on the chemical safety data sheet:

- name of chemical product and company identification, including trade or common name;
- information on the composition of ingredients;
- name and address of supplier or manufacturer;
- hazard identification;
- first-aid measures;
- fire-fighting measures;
- accidental release measures;
- handling and storage;
- exposure controls/personal protection;
- physical and chemical properties;
- stability and reactivity;
- toxicological information;
- ecological information;
- disposal considerations;
- transport information;
- regulatory information;
- other information (including date of preparation of chemical safety data sheet).

If chemical safety data sheets are not available, they should be immediately obtained from the supplier.

From the knowledge of the industrial process and the information on the chemical safety data sheet, a careful analysis needs to be carried out by management to determine chemical compatibility, and procedures for storage, transfer, handling, use and disposal. Such elements on the chemical safety data sheet as chemical and physical properties, stability and reactivity, and toxicological information are essential information for carrying out the analysis and planning appropriate control strategies. A complete set of chemical safety data sheets should be kept on file with the safety officer, occupational health service and in-plant fire brigade, if this exists, and should be immediately available to first-aid teams. When an emergency occurs and a worker is exposed to a chemical, the chemical safety data sheet should be brought to the doctor or medical facility to assist in rapid identification and treatment.

The information provided on chemical safety data sheets should also serve as a basis for the preparation of oral and written instructions to workers, and for the training of workers and supervisors in the safe use of specific chemicals. This training should include instructions for workers on how to obtain and use the information provided on the chemical safety data sheet.

Discussion questions:

1. Are chemical safety data sheets available to you for every chemical in your workplace? If not, how can you gain access to them?
2. Outline the way in which instructions for using chemicals safely are presented to workers and supervisors. Is this method effective?

4.2.4. Safe storage

If a hazardous chemical cannot be replaced by a less hazardous one, the quantity of the chemical at or near the workstation should be reduced to that required for daily use (one shift). The remainder should be kept in a safe chemical storage area.

The following general rules should be followed to ensure safety in the storage of chemicals:

- non-compatible substances should not be stored together (for example, the storage of acids near a cyanide compound could result in a spill and the generation of deadly hydrogen cyanide gas);
- storage of chemicals near an incompatible process should be avoided;
- chemical containers in the storage area should not be leaking, rusty or damaged, and must be properly stacked;
- there should be adequate ventilation to ensure that any leakage of hazardous vapours will be sufficiently diluted and eliminated.

For chemicals that pose a risk of fire or explosion, the following additional rules should be observed:

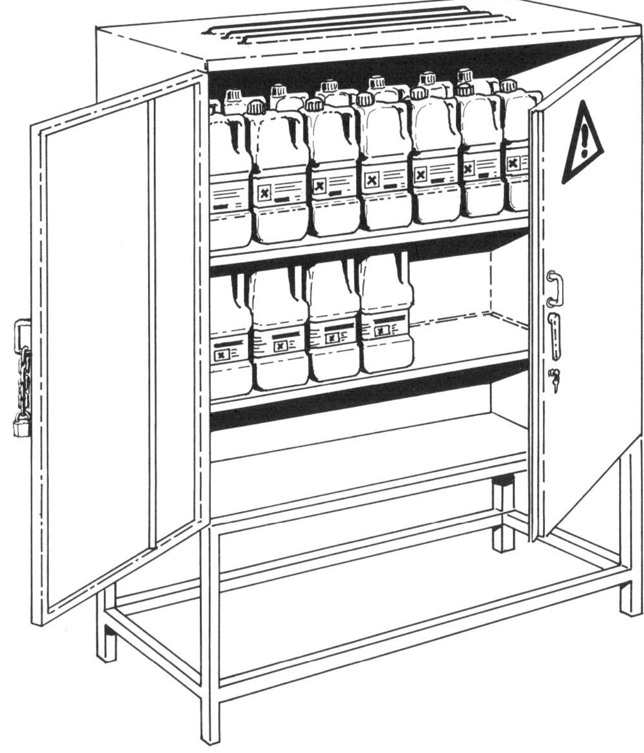

Figure 36. Flammable chemicals should be stored in cool, well-ventilated areas away from possible sources of ignition

- chemicals should be stored in cool, well-ventilated areas away from possible sources of ignition (figure 36);
- the storage facility should be separated from the plant and living quarters, and away from sources of drinking-water;
- an automatic fire protection system, such as a sprinkler or deluge system (where water-reactive chemicals are not present) should be provided;
- the plant should incorporate automatically closing fire doors, an alarm system, and a dyked area to prevent run-off after a fire;
- there should be easy access to fire vehicles;
- electrical circuitry should be explosion-proof, adequately fused to prevent overloading;
- wiring, switchboxes and fixtures should be protected from accidental damage due to the movement of drums, pallets and forklift trucks;
- to avoid possible ignition by static electricity, all drums positioned for transfer should be earthed and bonded;
- no sources of radiant heat should be present, and open flames from welding and smoking should be prohibited;
- forklift trucks equipped with internal combustion engines should be prohibited from the area;
- only quantities necessary for the operation of the plant should be stored.

Discussion questions:

1. Describe the special precautions that are taken in your plant for the storage of hazardous chemicals.
2. Describe how you can obtain further information about the storage of hazardous chemicals.

4.2.5. Procedures for safe transfer

Chemicals may be transferred to or from work areas through pipelines or conveyors, or by using forklift trucks, trolleys or wheelbarrows.

If chemicals are transported through pipelines, care must be taken to ensure that valves and flanges are intact and do not leak. If conveyors are used, the spread of hazardous dusts can be avoided if the conveyors and their transfer points are covered. If chemicals are transported at high speed and pressure through manifold systems, care must also be taken to avoid a build-up of heat, thus creating the risk of fire or explosion.

Containers for flammable liquids should be specially constructed with spring-loaded caps and flame arresters in their spouts.

The transfer of flammable liquids should only be carried out in well-ventilated areas with the containers earthed and bonded (figure 37).

If chemicals are transported by forklift truck, clearly marked passageways of an adequate width can reduce the possibility of collision and spillage.

Discussion questions:

1. Outline the procedures necessary in transferring a flammable liquid from a drum to a container.

Figure 37. A specially designed container for the transfer of small quantities of flammable liquids

2. Outline the safety precautions that should be taken when transferring chemicals by forklift truck.

4.2.6. Safe handling and use

As discussed in Chapter 2, there are three principal ways in which chemicals can enter the body, through absorption, inhalation and ingestion. In the workplace the most common route of entry is inhalation, followed by skin absorption.

To be inhaled, chemicals must be airborne in the form of dusts, vapours, mists or fumes. Dust is frequently formed during grinding, crushing, cutting, drilling or breaking operations. Vapours are generated by heated liquids or solids. Mists are produced from spraying, electroplating or boiling operations. Fumes arise from molten metals during welding or casting operations.

Skin absorption usually occurs when liquid chemicals are handled. Splashing of liquids on to exposed skin or clothing is the most common mode of contact. This may happen during operations such as dipping parts into degreasing tanks, using cutting oil during machine operations or transferring liquids.

Often these same processes pose a risk of fire or explosion due to the nature of the chemical involved. It is essential to control the sources of heat in order to reduce the risk if a substitute chemical is not available.

Certain precautions should be taken before handling or using chemicals:

- read and understand the instructions on the label and the chemical safety data sheet, and any other information provided with the chemical, relevant equipment and personal protective equipment;
- ensure that the user of the chemical has received effective training in the use of the chemical and the precautions to be observed;
- ensure that protective measures such as local ventilation or shielding are present and functioning properly;
- control the area where the chemical will be used for hazards that may pose a risk (such as an open flame or sources of fuel if a flammable liquid or gas is to be used) and remove the risk before using the chemical;
- check that the protective clothing and other safety equipment, including respirators if required, are complete, in good repair and of the correct quantity;
- ensure that any emergency equipment that may be necessary is readily available and in good working order.

During the handling and use of hazardous chemicals, prevention of exposure can best be achieved by following the principles of control, which are repeated here:

- elimination or substitution;
- enclosure or isolation;
- ventilation;
- the provision of personal protective equipment.

Discussion questions:

1. What precautions do you take before a hazardous chemical is used at your workstation?
2. How can you ensure that the appropriate protective measures are in working order before using a chemical substance? What do you do if you find that they are not in working order?

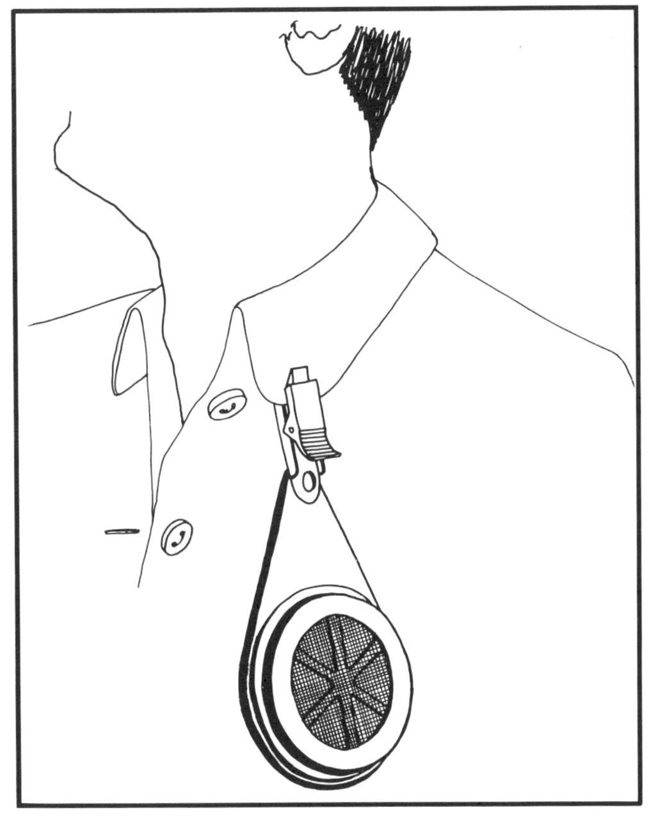

Figure 38. A personal monitoring device

4.2.7. Housekeeping

Good housekeeping plays an important role in the control of chemical hazards. Dust on workbenches, floors or ledges should be cleaned regularly by vacuuming rather than by compressed air or sweeping. Spilled liquids should be deposited in airtight receptacles and removed daily from the work area. Chemicals stored in damaged or leaking containers should be transferred to sound ones, and the damaged container disposed of accordingly.

4.2.8. Disposal routines

All manufacturing processes generate a certain amount of waste. Improper disposal of hazardous waste not only poses a health hazard to workers, and a potential fire and explosion risk, but also endangers the environment and people living in close proximity to the plant.

All waste products should be stored in properly labelled, specially designed waste containers. Any empty containers or bags previously containing toxic or flammable substances should be disposed of in such containers.

Figure 39.
Periodic medical examinations help detect early symptoms of occupational disease

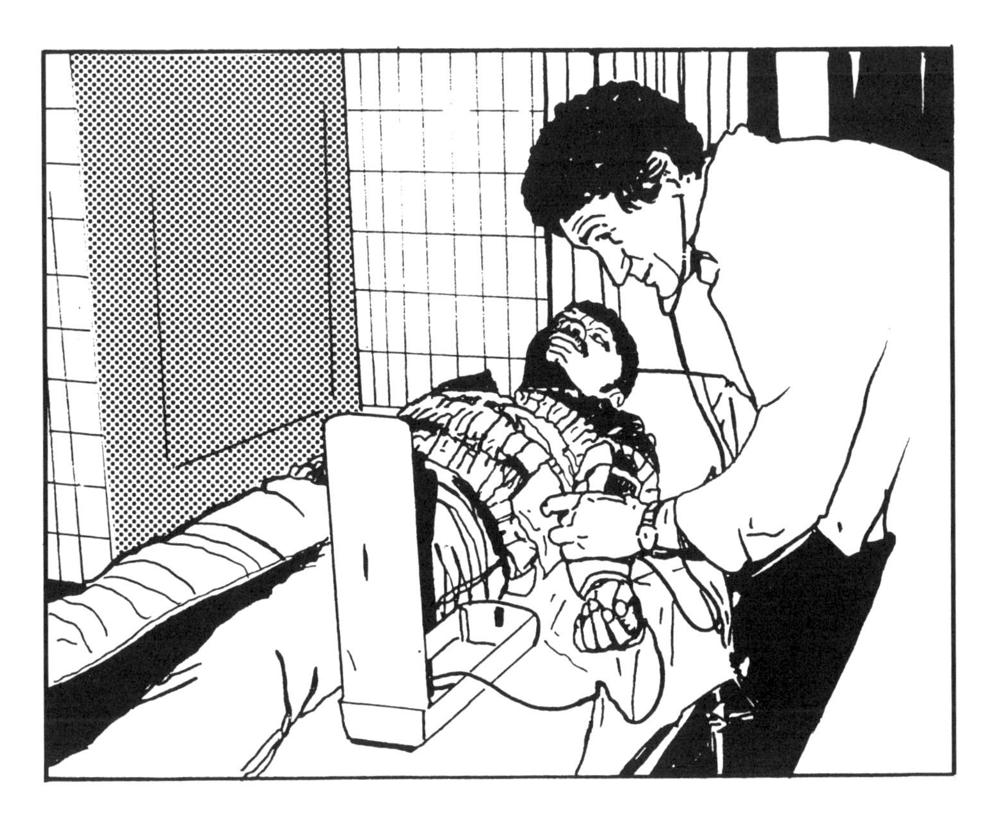

Figure 40.

Training and education provide workers with the knowledge and skills to prevent unnecessary exposure to hazardous chemicals

There should be an established written procedure for the disposal of toxic and hazardous waste. The safety of the workers handling the hazardous waste should also be ensured through appropriate control measures.

4.2.9. Monitoring of exposure

A monitoring programme in the workplace includes the taking of air samples to determine the concentration of chemicals in the atmosphere. These chemicals may be in the form of dusts, vapours, gases or fumes.

The air is sampled either by attaching a personal monitoring device within the breathing zone of the worker (figure 38) or by placing an air-sampling device in specific areas of the workplace.

The sampling can be done over either a short or long period of time. The analysis of the results will indicate the concentration of a specific chemical or other air contaminants that existed at the time of sampling. This concentration will in turn be compared either with an exposure limit value issued by the competent national authority or with any other widely accepted set of exposure limit values. Once a problem is identified, control measures should be implemented to reduce the exposure of the worker.

4.2.10. Medical surveillance

Medical surveillance includes pre-placement and periodic medical examinations (figure 39). Pre-placement examinations provide an opportunity to detect hypersusceptible workers and accordingly assign them to jobs or workplaces where their health will not be at risk. Periodic examinations help detect early symptoms of occupational diseases, and also verify whether control measures are operating efficiently.

4.2.11. Record-keeping

All environmental and medical surveillance records should be kept and maintained in good order. Some of the diseases caused by chemicals have long latency periods. These records will therefore be useful in the future to help medical practitioners make diagnoses for compensation purposes, and to provide valuable information for epidemiological studies which will contribute to the further understanding of the health hazards of chemicals.

4.2.12. Training and education

Training and education play an important role in the control of chemical hazards (figure 40). People who work with chemicals should be instructed in the possible hazards caused by the chemicals, safe working procedures, the care and use of protective equipment, and emergency and first-aid measures (see section 5.4 on first aid).

Workers should be trained to identify when control measures fail and to interpret the labels and hazard information provided for the chemicals. Training is essential for new workers, while existing workers should undergo periodic refresher courses.

Further information on training is given in sections 6.2.5.2 and 6.2.5.7.

Suggested further reading

ILO: *Encyclopaedia of occupational safety and health*, 2 vols. (Geneva, 3rd ed., 1983).

Joint Industrial Safety Council: *Safety, health and working conditions* (Stockholm, 1987).

National Safety Council: *Accident prevention manual for industrial operations*, 2 vols. (Chicago, Illinois, 9th ed., 1988).

5. Chemical emergency procedures

Throughout this manual we have stressed the *prevention* of disease, injury, fires and explosions. Realistically, however, it is not possible to prevent every accident from occurring.

Everyone who comes into contact with hazardous chemicals must not only be aware of preventive measures but must also know about emergency procedures, which may in fact prevent a minor incident from becoming a major catastrophe. These procedures include first-aid measures, fire-extinguishing techniques, and spill and leak procedures. Appropriate action in the first few seconds could prevent a disaster.

As with preventive measures, the key to emergency procedures lies with a sound knowledge of the chemical substances used in the plant and rapid access to sources of information. Chemical safety data sheets provide a great deal of information on first-aid measures, fire-fighting techniques, and spill and leak procedures. The label affixed to the chemical may also be an invaluable source of information during an emergency.

> **Remember:**
> *Knowledge about the chemical, sources of information and emergency measures are best obtained before an emergency occurs.*

It may also happen that chemicals that are stored together may accidentally mix during an emergency, forming a new substance with totally different characteristics. The plant chemist or industrial hygienist should be able to provide you with advice about the appropriate storage of chemicals, in order to keep non-compatible chemical substances away from each other.

If the possibility still exists that substances might mix, the plant chemist or industrial hygienist are the best people to guide you on the most reasonable course of action in an emergency.

5.1. The emergency plan

It is essential that every workplace should have an emergency plan. The plan should lay down the following procedures:

- the evacuation of workers, including a system of accounting for all workers outside the building;
- methods of notifying outside assistance such as medical, rescue, fire or environmental protection specialists, as necessary;
- the role of various plant officials during an emergency;
- the role of selected workers;
- the location, use and maintenance of all emergency equipment in the plant.

Everyone in the plant should be informed of the emergency plan and should understand it in detail. The plan should describe clear and unobstructed emergency exits, a functioning and frequently tested alarm system, and training in evacuation for all workers. It should also detail procedures for the immediate evacuation of disabled workers who may need assistance in reaching emergency exits.

There should be meeting places outside the plant so that each worker can be accounted for after evacuation. These predetermined meeting areas should be safe from the possible escalation of the emergency. The emergency plan should outline the structure of the first-aid organization within the plant, as well as procedures to obtain more specialized medical care when and as necessary. The role of all plant personnel (including workers, supervisors and managers) during a first-aid situation should be described. The location of all emergency first-aid equipment, including emergency showers, eye-wash stations, first-aid kits and stretchers, should also be mapped out.

The plan should address the organization of the internal capability to fight small fires within the plant. As with first aid, the role of all plant personnel in a fire emergency must be described, even if it only details the procedures for rapid evacuation. The location of all fire-fighting equipment such as sand buckets, hoses and extinguishers, as well as automatic fire-fighting systems, should be described with specific guidance as to who should fight a chemical fire and when.

A chemical leak or spill can have disastrous consequences when the situation is not dealt with rapidly. The emergency plan should specify who will be involved in controlling the leak or managing the spill. Again, any specific material or equipment must be described. If an emerency spill or leak team is foreseen, the plan should detail its structure and responsibilities.

Emergency plans should be developed in conjunction with local medical, fire, law enforcement and civil defence authorities, as well as neighbouring plants.

> **Remember:**
> 1. *Every workplace should have an emergency plan.*
> 2. *The plan should cover emergency exits and an alarm system for evacuation.*
> 3. *It should outline the duties and responsibilities for first aid and fire-fighting within the organization.*

Discussion questions:

1. Does your plant have an emergency plan? If not, how can you assist in setting one up?
2. How can you test the effectiveness of the plan?

5.2. Emergency teams

There is much to be said for setting up and maintaining emergency teams designed to deal with the three types of problem encountered in a chemical emergency, i.e. first aid, fire, or spill and leak. However, it must be recognized that a team of only two or three workers cannot deal with all the elements of an in-plant chemical emergency. A small number of workers cannot provide first aid, fight a fire and clean up a spill at the same time.

Waiting for an emergency team or external emergency resources to respond can turn a minor incident into a major catastrophe. Therefore every worker should have enough basic training in the above procedures to allow for action in a chemical emergency.

5.3. Evacuation

Every workstation should have at least two clearly marked, unobstructed emergency exits (figure 41). Provisions should be made to ensure lighting even during a power failure. If the evacuation route requires personal protective equipment because of the potential presence of a toxic chemical, the equipment should be maintained in a constant state of readiness, inspected and readily available without locks or

Figure 41.
Every workstation should have clearly marked, unobstructed emergency exits

Figure 42. When someone is injured, as shown here, first aid can make a great difference

keys. All workers should be adequately trained – and frequently retrained – in its use.

Remember:
Every workstation must have clear and unobstructed emergency exits.

5.4. First aid

5.4.1. The organization of a first-aid service

Some form of first-aid service is essential to every workplace (figure 42). Additional requirements apply when hazardous substances are present. When a first-aid service is being considered, a number of aspects need to be evaluated:

- the nature, quantity and hazards of the substances present;
- the availability of trained first-aid/medical personnel;
- the proximity of the nearest medical facility;
- the availability of transport to the nearest medical facility;
- the ability to communicate in order to obtain outside help, such as telephone or two-way radio;
- the emergency equipment within the plant, such as emergency showers and eye-wash stations;
- the training of workers in basic emergency procedures.

5.4.2. First aid for persons who come into contact with chemicals

It is essential that the provision of first aid to the victim of an accident puts neither the person providing the first aid nor the victim at further risk. In particular, when a worker needs to be rescued from a chemical area or is grossly contaminated, the rescuer needs to take precautions so that he or she does not become a victim:

- if a worker is overcome by gas or fumes, the rescuer needs suitable breathing apparatus (usually self-contained) before entering the danger area;
- if the skin or clothing of the worker is highly contaminated, the injured worker should be flushed with water as the clothing is removed;

Figure 43. Remove the casualty to an uncontaminated place

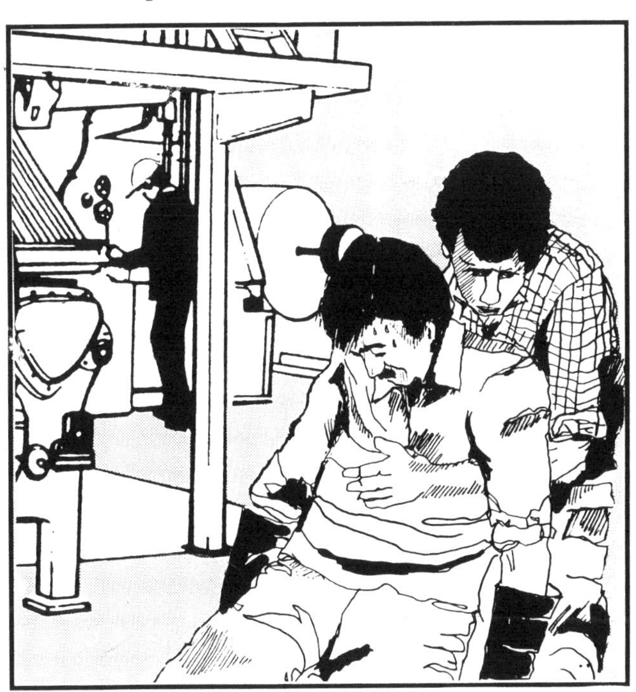

Figure 44.
A casualty placed in the recovery position

- when rescue is needed from a confined space, sump, pit or underground area, an alarm should be given to call additional help. Entering a dangerous situation alone to help someone else may lead to two victims;
- the injured worker should be carefully removed from the dangerous environment to a safe area (figure 43) and placed in the recovery position (figure 44). If the worker is unconscious, he or she may have to be dragged on a blanket or pulled head first by grasping the clothes.

> **Remember:**
> *Prior to administering first aid, the rescuer should ensure that the accident victim is in a safe environment. The victim should be carefully moved if necessary.*

There are several very urgent priorities in providing first aid for a chemically injured worker:

- if there is no pulse, and a person trained in cardio-pulmonary resuscitation (CPR) is available, then CPR should be administered (figure 45);
- most chemical injuries will be chemical burns of the skin or eyes. If the worker has had a hazardous substance splashed on the skin or in the eyes, the chemical should be flushed away with large volumes of water for at least ten minutes, unless otherwise indicated (figure 46);
- if clothing is contaminated, it should be immediately removed or cut off, all the while flushing with large amounts of water;
- if an emergency shower is available, the injured worker should be placed in the shower and *all contaminated clothing should be removed* under running water.

> **Remember:**
> *Both mouth-to-mouth resuscitation and flushing of contaminated skin or eyes should be carried out while avoiding further injury.*

If a worker has accidently swallowed a chemical substance, the first-aid treatment will depend on the nature of the substance. With most substances, if the worker is conscious, it is essential to make him or her vomit. With other

Figure 45. Cardio-pulmonary resuscitation

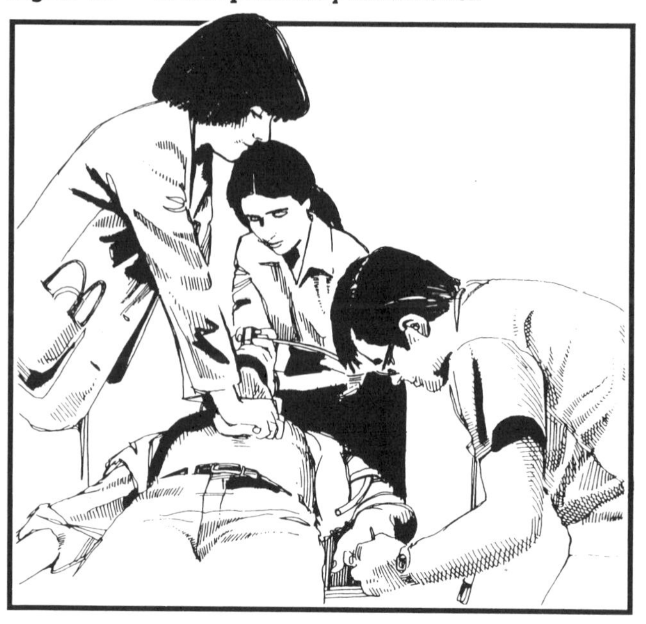

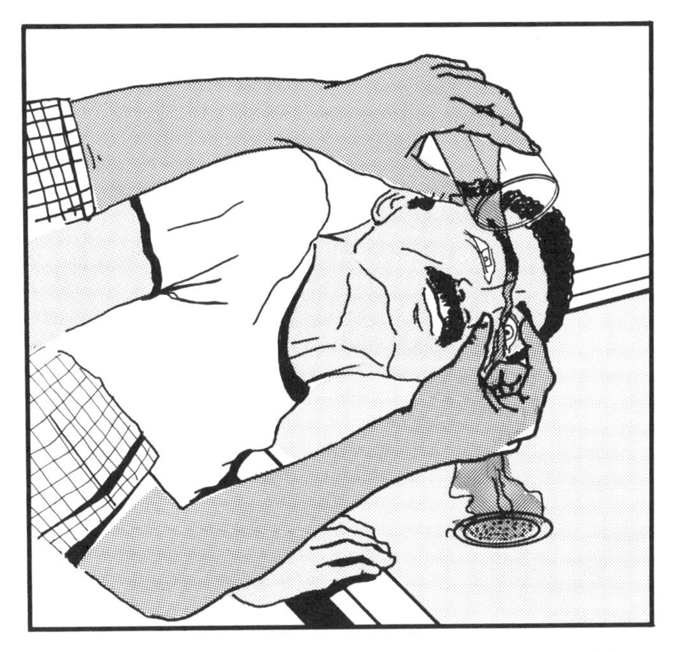

Figure 46. If chemicals enter the eyes, wash thoroughly with clear running water

substances such as petroleum products or organic solvents, it is dangerous to vomit. It is also dangerous to make an unconscious worker vomit. The label on the container or chemical safety data sheet may be able to provide the necessary guidance.

Any injured worker who is drowsy or unconscious must be rushed to a medical facility or hospital as quickly as possible, and accompanied by someone who is ready to give mouth-to-mouth resuscitation if it becomes necessary.

Even if the worker appears to be all right, seek medical assistance as rapidly as possible.

Discussion questions:

1. Can you identify first-aid trained workers in your plant?
2. How can you best ensure that they have adequate knowledge of hazardous chemicals?

5.4.3. Role of poison information centres

A fair number of countries now have access to poison information centres which have been set up in response to the growing need for medical advice about chemical and pharmaceutical products.

The role of each centre is essentially to provide a support service to doctors, emergency services and other health workers who are called upon to treat cases of acute poisoning. The service operates by reference to an extensive computerized index of substances which describes their toxicity, diagnosis and treatment. Advice is almost always in response to a telephone call or other means of obtaining an urgent reply. In some countries centres are operated 24 hours a day throughout the year.

Centres may also provide further services such as:

- providing antidotes for poisons, particularly those which are not widely available;
- coordinating the activities of medical experts to treat particular cases;
- providing a laboratory service to analyse blood or other samples for poisons;
- identifying trends from all inquiries to determine causes of poisonings which point towards the need for a particular solution, such as improvements to labelling or packaging;
- analysing inquiries on behalf of government or manufacturers in respect of particular products;
- educating and informing others about their work and on improvements to enable better diagnosis and treatment.

Employers and managers in enterprises where chemicals are used should establish contacts with poison information centres, where they exist. Such contacts have proved to be vital in saving lives in cases of chemical poisoning.

5.5. Fighting fire

5.5.1. The pre-fire plan

Should a small or a major fire occur, it is essential that everyone be clearly aware of their roles and responsibilities (figure 47). It is also very important that information be available describing

Figure 47. Fires in the workplace can happen and should be planned for in advance

the internal fire-protection equipment, procedures for evacuation of plant personnel, when and when not to fight fire, and how the location of specific chemicals (in storage or in process) might affect the safe fighting of fire.

The pre-fire plan needs to be flexible, and will change as chemicals are added to or removed from the workplace. It will also alter as buildings, industrial processes and fire-protection equipment are put into service, removed from service or changed in any way.

For example, the role of an in-plant fire brigade should always be clearly defined in the pre-fire plan. This role will be very different if an automatic sprinkler system is in place and functioning. The brigade's task will be to assure the safe operation of the sprinkler system and support immediate evacuation from the affected area. The in-plant fire brigade's role may also change depending on the time it takes for a municipal fire service to arrive at the scene of a fire. If the response time is long, it may be necessary to develop a higher level of response within the plant.

As a minimum, the pre-fire plan should describe the following:

- information about the chemical fire risks within the plant, including the potential for applying certain fire-extinguishing agents to certain chemicals and the need for personal protective equipment (much of this information may be available from chemical safety data sheets);
- information about the capacity of the municipal fire department to assist the plant in dealing with fire emergencies involving chemicals within the plant;
- information about the in-plant fire brigade, including its structure, training programme, equipment and capacity to deal with fires involving chemicals;
- the interrelationship between the in-plant fire brigade and the municipal fire department;
- the fire-protection equipment available within the plant, including automatic sprinklers, fire extinguishers, hoses, etc. (figure 48);
- the fire-alarm system;
- the fire-escape and evacuation plan;
- the frequency and type of fire-drills needed for the plant.

Most importantly, the pre-fire plan should list in detail the actions to be taken by all workers if there is a fire. The plan must state who is going to do what, when and where. *Remember, the safety of the workers is of the highest priority.*

5.5.2. The organization of in-plant fire-fighting services

Depending on the size of the plant, the number of workers and the nearest sources of outside

Figure 48. Fire-alarms and equipment, and emergency exits, should be part of the pre-fire plan

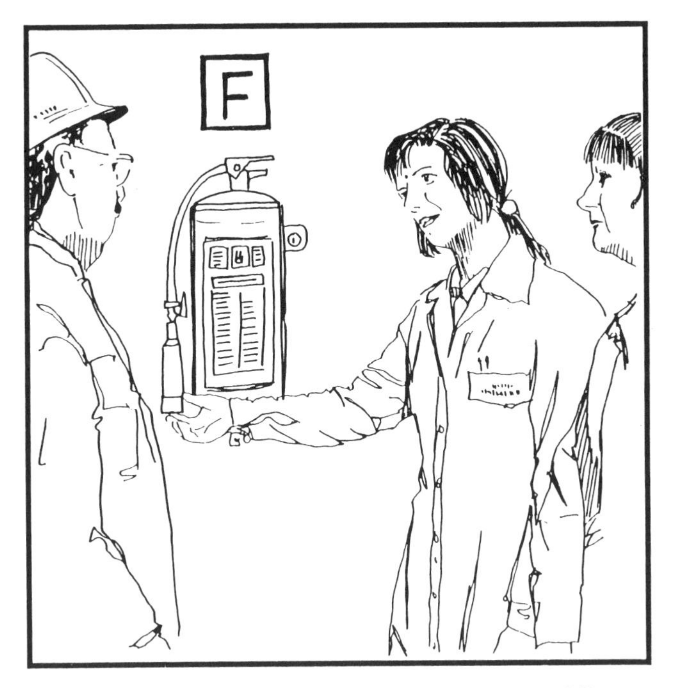

Figure 49. It is important to select the right type of fire extinguisher

assistance, it may be worth while for a plant or a group of plants to organize an in-plant or shared industrial fire brigade.

If an in-plant fire brigade is trained, equipped and prepared to respond to chemical emergencies, the amount of time required to bring a chemical fire under control can be greatly reduced and financial losses considerably lessened.

If the plant fire brigade is to address chemical emergencies, it must consider the following:

- Is there adequate information on the behaviour of the chemicals used and produced in the plant to predict how they will react when they burn?
- Will the chemicals produce toxic or explosive gases as they are heated?
- Is there adequate training available for the fire brigade to fight a chemical fire safely?
- Do the members of the fire brigade have adequate personal protective equipment to protect them during fire-fighting operations?
- To whom will the fire brigade turn if it reaches a point where it can no longer control a fire?

It is strongly recommended that the plant owner or manager should establish direct contact with local or municipal fire services when considering whether or not to have a plant fire brigade address chemical emergencies.

5.5.3. Automatic fire protection

If a fire activates an automatic fire-protection system, the worker or in-plant fire brigade should not try to inhibit or interfere with the operation of the system. Many times has a minor fire become a major one because someone with good intentions prematurely shut down the automatic fire-protection system.

5.5.4. Selection of extinguishers

Portable fire extinguishers are very effective for the control of small chemical fires before they can assume large, uncontrollable proportions. It is imperative that the right fire extinguisher is chosen for a given chemical hazard (figure 49). The chemical safety data sheet will normally indicate the best extinguishing agent for a fire involving a given chemical substance. The type of extinguisher placed around the workstation should reflect the fire risk involved.

The choice of fire extinguishers for combined substances should only be considered after consultation with a person who is competent in the control of chemical fires.

Table 3. Types of fire extinguisher: Fire-fighting action and associated risks

Type	Action	Risk
Pressurized water	Cools fuel rapidly	Conducts electricity Reacts with some chemicals
Carbon dioxide	Removes oxygen	Displaces oxygen when used in confined spaces
Dry chemical	Interferes with the combustion process	When used in confined spaces, greatly limits visibility

It must also be considered that certain types of fire extinguisher have different risks associated with them. Table 3 indicates the fire-fighting action and associated risk of various types of extinguisher.

5.5.5. The decision to fight fire

The first priority in the control of a fire is to support efforts to evacuate plant personnel. A decision by the worker or a supervisor to fight a fire should be made only if it is considered possible to do so without any threat to life. Excessive heat, the risk of explosion, and the possibility of being overcome by a lack of breathable air or the risk of being trapped by a spreading fire should all be considered in making this decision.

Remember:

If the decision is made to fight a chemical fire the following points should be taken into consideration:

- *the person fighting the fire should never do so alone;*
- *there should always be a clear, unobstructed exit, clear of fire so that rapid escape can be safely accomplished;*
- *the appropriate extinguishing agent must be chosen to control the fire effectively and safely;*
- *once the fire has been extinguished, do not replace the hoses and extinguishers that have been used so that they can be inspected and recharged.*

5.6. Spill and leak procedures

If there is a spill of a hazardous chemical from a container, or a leak from a tank or other source such as a pipe or vessel, certain procedures should be followed. As with other emergencies, these procedures should always be planned in advance and documented in an emergency plan.

A key to the successful control of a spill or a leak is the knowledge of the properties and behaviour of the chemical substances involved (figure 50). Again, the best source of information is the chemical safety data sheet for individual chemicals, or the plant chemist or occupational hygienist for spills that involve more than one chemical.

It is also essential that the plant personnel dealing with the spill or leak should immediately be able to judge whether or not the situation can be handled from within the plant or whether outside assistance is needed.

Depending on the size and nature of the spill or leak, and the hazardous chemical involved, the following steps must be taken:

1. Evacuate any non-essential personnel to an area safe from any possible harm and provide emergency first aid if called for.
2. If the chemical is flammable or combustible, reduce the risk of fire or explosion by extinguishing any open flames and any other sources of heat or ignition.
3. Evaluate the extent of the situation and the ability of plant personnel to deal with it. If necessary call in outside assistance.
4. Assume that this is an abnormal situation. Although personal protective equipment may not be necessary in the day-to-day handling or use of the chemical, a spill or a leak may go beyond the operational controls that normally

Figure 50. Use correct procedures for dealing with a spill or leak

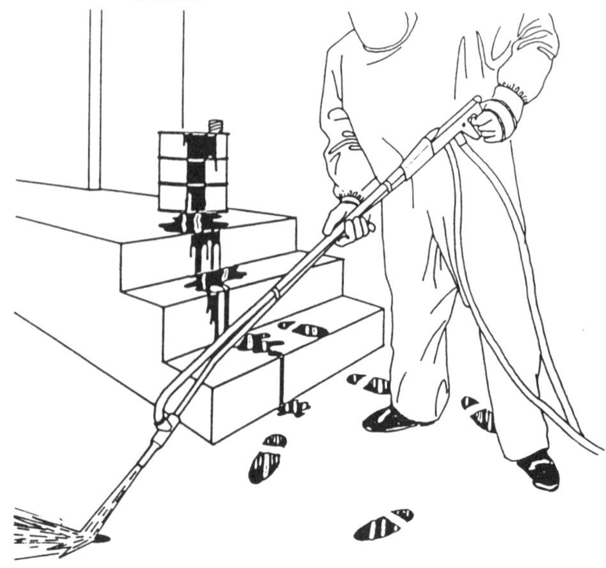

apply. From the chemical safety data sheet determine in advance the personal protective equipment necessary to deal safely with the situation.

5. Eliminate the further spread of the chemical involved by controlling it at its source if possible. This may be done by closing a valve, sealing a tank or rerouteing a process. These actions should be performed by a competent person knowledgeable about the process in order to avoid any further conditions that could lead to additional risks.
6. Attempt to contain the spill or leak by dyking and absorption. If appropriate, the chemical should be either sealed in containers or neutralized.
7. Once the chemical is safely stored or neutralized, the area of the spill or leak must be decontaminated, inspected and monitored if appropriate by qualified personnel.
8. If the area is found to be safe, normal activities can resume.

Remember:

1. *Information on methods for dealing with spills and leaks can be found on the chemical safety data sheet.*
2. *The first priority must be the evacuation of non-essential personnel.*

Discussion questions:

1. Describe the spill or leak procedures for the chemicals that you use at your workstation.
2. If these procedures are not formally documented, what measures can you take to do so?
3. How is the evaluation made at your workplace to determine whether outside help is needed to deal with a spill or leak?

Suggested further reading

American Red Cross: *Advanced first aid and emergency care* (Washington, DC, 1979).

David T. Gold. *Fire brigade training manual* (Quincy, Massachusetts, National Fire Protection Association, 1982).

ILO: *The organisation of first aid at the workplace* (Occupational Safety and Health Series No. 63 (Geneva, 1989).

National Fire Protection Association: *Fire protection manual* (Quincy, Massachusetts, 17th ed., 1991).

6. Management of a chemical control programme

The ultimate responsibility to use chemicals safely in an enterprise rests with management. Management has the authority and the resources to develop and carry out programmes spelling out methods and procedures on safety and health in the use of chemicals at work (figure 51). To be effective, the programme for managing the safe use of chemicals should receive a priority similar to other programmes within the enterprise such as production, marketing, maintenance and quality control. The measure of success will materialize in the form of a better working environment, a reduction in workplace accidents and diseases, a healthier workforce and a reduction in waste. Together, these elements will lead to increased productivity and profits for the enterprise.

This chapter describes the steps to be taken in managing a chemical control programme.

Remember:

There are three ideas underlying a programme for the safe management of chemicals:

1. *Management should know all the chemicals being used in the enterprise, including their quantities and their associated hazards.*
2. *Workers should know the hazards associated with the chemicals they work with and should be trained in the necessary precautions.*
3. *Workplaces should be designed or adapted to the needs of the worker instead of attempting to adapt the worker to the workplace.*

Figure 51. Activities comprising the management of a chemical control programme

6.1. Setting goals

In order to control any programme activity successfully, a clear policy for the safe use of chemicals is required. This is a broad statement of goals by top management which is made known to managers, workers and others who deal with the enterprise so that the way they do their jobs and carry out decisions is guided by the policy. This general principle can be incorporated into a policy statement for a comprehensive control programme.

A policy statement on the management of chemicals might include the following commitment:

- safe procedures and practices will be established for the transport, use and disposal of hazardous chemicals;
- management will ensure that the workers have the right to be fully informed on the hazards of chemicals and to be thoroughly trained in their safe handling;
- before any chemical is brought into the enterprise, information on that chemical should be provided by the supplier, manufacturer or importer.

To complement the policy statement, the enterprise should list its main priorities that would enable it to set goals. For example, before a hazardous chemical is used, the enterprise may require a full investigation into the degree of risks it poses, taking into account the economic and operational impact of substitution. Another priority may require that exposure levels of a hazardous chemical should be at the lowest practicable level if it is found that the use of such a chemical is essential for the process. Since these listed priorities provide the means to set goals, they should be as wide-ranging as possible.

Discussion questions:

1. Describe your enterprise's policy for a chemical control programme.
2. Identify who is directly responsible for the implementation of the policy within the enterprise.

Remember:
Every enterprise should have a clear policy on the safe management of chemicals.

6.2. Setting up a programme

6.2.1. Cooperation at the workplace

To coordinate and plan activities on safety and health in the use of chemicals, it is important to identify a group of people who take the initiative of establishing a plan and monitoring its implementation. Depending on the size of the enterprise, the size of the group could range from two people (a representative of management and a workers' representative) to a working team. Ideally, the group should consist of a representative from management with authority, a safety officer and an industrial hygienist (to provide the professional backing), as well as representatives from among the workers using chemicals and having responsibility for storage.

For small organizations which lack professional support, outside assistance may be sought, especially during the initial stages. This assistance might come from a consultant, local trade associations or a government agency. It is essential that the group receives the full support of management. The group's activities should be closely guided by the strategies as reflected in the enterprise's policy.

Employers and workers have certain roles and responsibilities which should be reflected in the chemical safety programme.

Role of employers

Employers should:

- ensure that chemicals are stored safely, and that unauthorized access is prevented;
- ensure that workers are protected against accidents, injury and poisoning at work by:
  - (a) as far as possible, choosing chemicals that eliminate or minimize the risk;
  - (b) choosing the appropriate equipment and machinery for work with chemicals;
  - (c) making sure that all chemicals are correctly labelled, and that chemical safety data sheets are provided and made available to workers and their representatives;
  - (d) instructing all workers, particularly those who are new or functionally illiterate, about the hazards and the safety precautions;
  - (e) undertaking effective supervision of all tasks involving chemicals to ensure correct operation and prevent any hazards that may result from lack of knowledge or experience of workers;
  - (f) carrying out maintenance, repair and periodic inspection of equipment and machinery, and workplaces;
  - (g) complying with safety and health regulations and safe working practices;
  - (h) making arrangements to deal with emergencies.

Employers, in carrying out their duties, should cooperate with workers and their representatives with regard to safety and health in the use of chemicals at work.

Role of workers

Workers should cooperate with employers in carrying out their duties and should comply with all procedures and practices relating to safety and health in the use of chemicals at work. They should note and follow the instructions given by the manufacturer, supplier, employer or supervisor. Workers should take all reasonable steps to minimize risk to themselves or others, property or the environment. In addition, they should:

- use properly all devices provided for their protection or the protection of others;
- examine the equipment before beginning work and report forthwith to their immediate supervisor any situation which they believe could present a risk, and which they cannot properly deal with themselves.

Workers should also have the right to:

- draw the attention of their representatives or the employer to potential hazards arising from the use of chemicals at work;
- remove themselves from danger resulting from the use of chemicals when they have reasonable justification to believe that there is an imminent and serious risk to their safety and health;
- request alternative work, in the case of a health condition placing them at increased risk of harm from a hazardous chemical; in particular, women workers who are pregnant or breastfeeding should be able to request transfer to work that is not harmful to the unborn or nursing child;
- require adequate medical treatment and compensation for injuries and diseases.

Discussion questions:

1. Who is included in the group for planning and monitoring the safe management of chemicals?
2. What are the roles and responsibilities of management and workers in your enterprise concerning the safe use of chemicals?

6.2.2. Inventory of chemicals

The first task of the group responsible for the coordination and planning of safety in the use of chemicals is to make a master inventory of the chemicals in use. The inventory should include every chemical used, its physical state and the estimated quantity used per month. This inventory process can only be accomplished by inspecting every work section, raw materials depot and storage area. Methods used for the transport, handling, storage and disposal of chemicals should also be noted.

The purpose of establishing the inventory is to develop basic safety information for the safe use of all chemicals in the enterprise. Chemical safety data sheets will provide most of the necessary data. (If these are not immediately available, the supplier of the chemical should be contacted and requested to provide the information.) From this information, the committee should analyse the use of hazardous chemicals and consider substituting less hazardous ones. If substitution is not technically and economically feasible, other preventive measures should be exploited such as installing engineering controls and implementing safe procedures and practices in the use of chemicals.

Safety and health data should also be made known to all workers who come into contact with the chemicals through a training programme and a manual on safe work procedures and practices.

Remember:

Chemical safety data sheets will provide the basic safety information for chemicals in the workplace.

Discussion questions:

1. Has a master inventory been developed for every chemical within your enterprise?
2. Is there a policy to update the master inventory regularly?
3. Are chemical safety data sheets readily available for each chemical?

6.2.3. Purchasing procedure

The second task of the group is to become involved in establishing and monitoring a purchasing procedure based on the rationale that all chemicals which enter the workplace should be endorsed by the group, properly identified, classified and labelled. Through this procedure, the group could examine and monitor whether the purchase of new chemicals would be suitable for use. The procedure should also call for the enterprise to consult the relevant national authority on regulations to determine labelling requirements and include these requirements in the purchase specifications.

Finally, the procedure should include a regular review of chemicals currently in use to evaluate whether or not their use should be continued.

Discussion question:

Describe the purchasing procedure in your plant which ensures that all chemicals are properly identified, classified and labelled.

6.2.4. Receipt, identification, classification and labelling

The third task of the group is to collaborate with the purchasing department in taking the necessary steps when a chemical first enters the workplace. It must be ensured that each chemical is correctly identified, classified and labelled, with an up-to-date chemical safety data sheet on hand (Annex 4 offers guidance in this respect). These conditions should always be met before the chemical is placed in storage or put into use. This task can be facilitated if the enterprise's purchasing department spells out the requirements in the purchasing specifications.

Every container holding a chemical should be clearly marked with a label. The purpose of the label is to provide everyone who comes into contact with, or is working in close proximity to, the chemical with essential information regarding its identity, classification, the hazards it presents and the safety precautions to be observed. Supervisors and workers should cooperate with management to make sure that each chemical container and package has a properly prepared label.

Section 4.2.2 gives a complete list of information that should be included on the label.

Discussion:

Describe a method whereby all containers in the workplace holding chemicals can be clearly marked with the necessary information.

> **Remember:**
> *Every container for chemicals should carry an appropriate label. Unlabelled containers should not be used.*

6.2.5. Day-to-day management of chemicals: Control measures

There are certain concrete activities that management should undertake to ensure safety and health in the use of chemicals at the workplace. These should apply to all workstations within the enterprise:

1. Ensure that all chemicals are in appropriate containers with proper labels and chemical safety data sheets.
2. Provide information and instruction to all workers concerning safe use and storage.
3. Ensure cooperation for improved control.
4. Manage the provision, use and maintenance of personal protective equipment.
5. Develop, periodically evaluate, and carry out drills in emergency measures.
6. Establish and maintain exposure monitoring procedures, including medical surveillance.
7. Plan and implement training programmes.

Management should address the daily issues pertaining to the use of chemicals with a view to protecting workers from the risks arising from day-to-day work. Exposure can be eliminated or reduced by establishing safe procedures and practices, as already explained in this manual. The best method would be to replace hazardous chemicals with less hazardous ones. If this is not possible, the processes using the chemicals should be enclosed, or local ventilation systems installed and maintained regularly. Personal protective equipment should only be provided as a supplement to other operational control measures.

The remainder of this chapter will discuss in detail the control measures listed above.

Ensure that all chemicals are in appropriate containers with proper labels and chemical safety data sheets

Many systems are available for the classification and labelling of chemicals (Annex 4 describes the system used in the European Communities – EC). Classification varies from one country to another. The responsibility for the correct classification and labelling rests with the supplier.

In the working environment, chemicals are classified according to their potential hazards to workers. The criteria for classification should be based upon the chemical's inherent health and physical hazards and should include the following:

- toxic properties (both acute and chronic);
- chemical and physical characteristics (including flammable, explosive, oxidizing and dangerously reactive properties);
- corrosive and irritant properties;
- allergenic and sensitizing properties;
- carcinogenic effects;
- teratogenic and mutagenic effects, producing malformation of the foetus or genetic changes;
- effects on the reproductive system.

Figure 52. An example of classification of chemicals, and the corresponding danger symbols, as used in countries of the European Communities

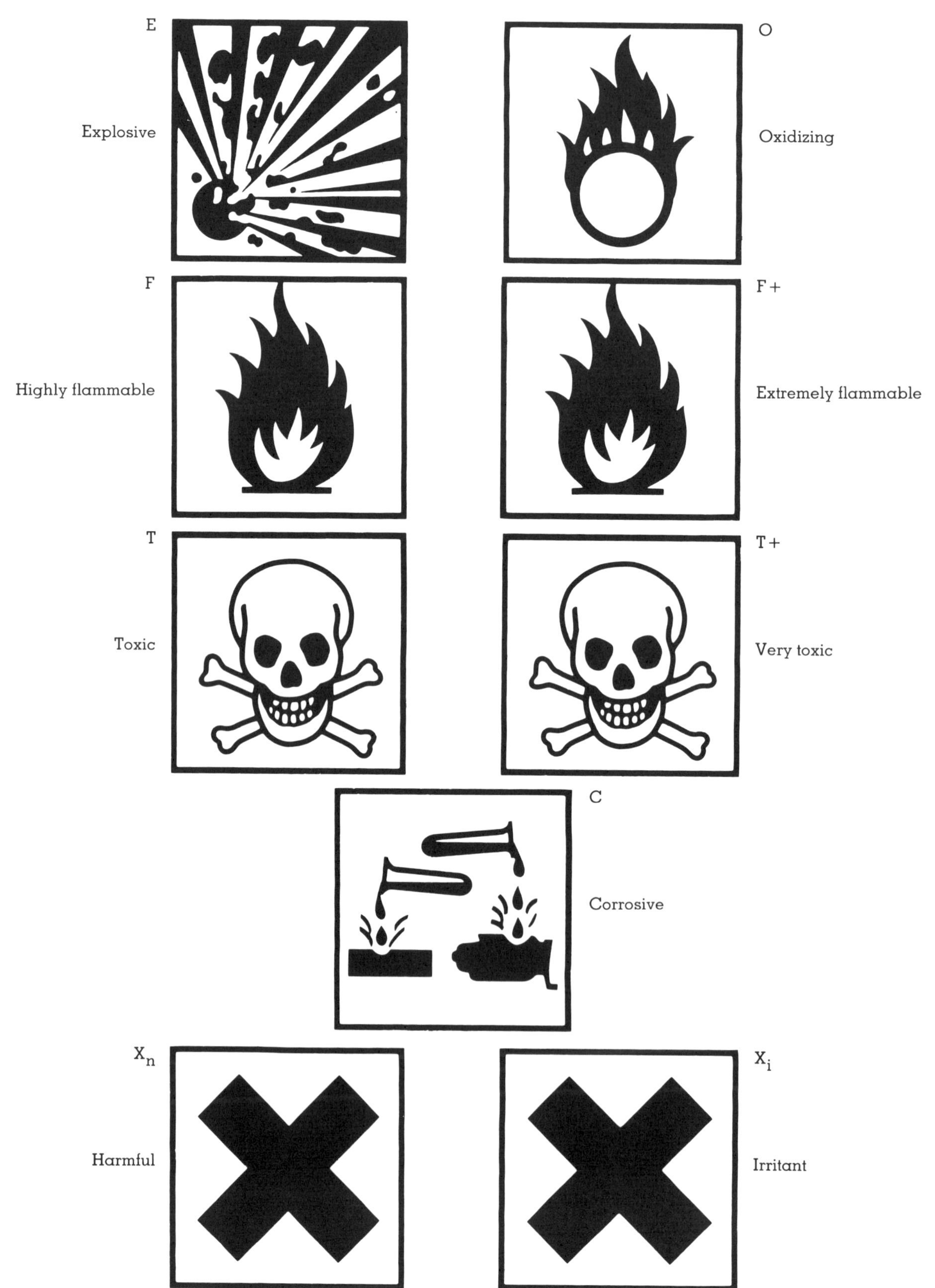

For example, a chemical can be classified as explosive, oxidizing, flammable, toxic, corrosive, irritating or harmful, along with appropriate danger symbols (figure 52 gives an example from the EC). The national authority should be consulted on the classification criteria for different types of chemicals.

The need for labelling of every chemical in the workplace and for chemical safety data sheets has already been mentioned in this manual.

At the enterprise, unlabelled chemicals that enter the workplace should not be handled until the supplier provides adequate information. Chemicals that have been dispensed into other containers should always be adequately relabelled. Many accidents have happened because chemical containers were not relabelled.

Discussion questions:

1. What danger symbols (if any) are used in your country to classify dangerous chemicals?
2. Describe the information that should be on the label and the chemical safety data sheet of a hazardous chemical.
3. Outline what steps can be taken to assure that every worker understands the information on the label and the chemical safety data sheet.

Provide information and instruction to all workers concerning safe use and storage

Management has the responsibility to ensure that:

- all workers are informed of the hazards associated with exposure to chemicals used in the workplace;
- workers are instructed how to obtain and use the information provided on labels and chemical safety data sheets (figure 53);
- the chemical safety data sheet, along with information specific to the workplace, is used as a basis for the preparation of instructions to workers;
- workers are trained on a regular basis in the practices and procedures to be followed to ensure safety and health in the use of chemicals at work;
- workers are also trained in the correct and effective use of control measures, in particular engineering control measures and measures for personal protection.

As arrangements for emergencies frequently involve workers, management has the responsibility to inform workers clearly of their role and to train them in the exact measures to be carried out during an emergency.

Proper storage is another important element in a programme for the safe management of chemicals. Management should consider factors such as:

- compatibility of chemicals;
- the properties and quantities of chemicals to be stored;
- the security of, and access to, the site;
- the nature and integrity of containers;
- the influence of environmental factors such as temperature and humidity;
- precautions against accidental toxic release and fire.

Figure 53. Workers should be informed on how to obtain data from chemical safety data sheets and labels

Different classes of chemicals require different types of storage. For example, flammable chemicals should not be stored in the proximity of oxidizing materials and the storage area should be cool, away from sources of ignition and well ventilated. Chemicals which react with water (lithium, sodium, potassium, calcium) should be stored in a dry, cool and well-ventilated area. Sprinkler systems should not be installed in this area.

Discussion questions:

1. Describe the special precautions necessary for the storage of flammable liquids.
2. How are workers trained to take special precautions when dealing with hazardous chemicals?

Ensure cooperation for improved control

An integral part of the management of safety and health in the use of chemicals is cooperation between employers and workers. It is essential that management should foster a spirit of cooperation, allowing for close collaboration in the application of measures to promote the safe use of chemicals at the workplace (figure 54).

Cooperation also means that safety rules and procedures should be followed and any dangerous situation arising from faulty or inoperative control or protective devices should be reported immediately to management. As a general rule, workers should carry out their tasks safely and in a manner which will not pose any danger to other workers.

If requested by the worker, management should be able to provide information on the health hazards of chemicals or other information such as exposure monitoring data or medical examination data for a particular worker.

Remember:
Cooperation is an essential element of a successful occupational safety and health programme.

During an inspection or investigation, the workers' representative should be allowed to participate if a request has been made to the management. Such cooperation will ensure the effectiveness and success of the chemical control programme.

Discussion questions:

1. How are workers represented in questions pertaining to hazardous chemicals?
2. Are safety representatives involved in inspections and investigations?

Manage the provision, use and maintenance of personal protective equipment

Where risks cannot be eliminated or hazardous processes made safe, it is up to management to provide and maintain the appropriate personal protective equipment, including clothing, at no cost to the worker. Management should also implement measures to ensure the appropriate use of this equipment.

Figure 54. Cooperation between employers and workers is essential in developing a programme to manage safety in the use of chemicals

Chemical safety data sheets are the key source of information to management and workers concerning the selection of appropriate personal protective equipment for the risks that a chemical may present. Based on this and other information from individuals such as industrial hygienists and chemists, or from laboratories or the competent national authority, a programme should be developed (figure 55). This should include:

- a written management policy statement on the use and maintenance of personal protective equipment;
- a method of providing workers with the proper type of personal protective equipment, with an adequate selection of sizes to fit all workers;
- a systematic method of posting areas and processes where personal protective equipment is required;
- a training programme which addresses the hazards, and the methods of protection, selection and use of personal protective equipment, as well as maintenance and repair.

Figure 55. Management should develop and implement a programme for the provision and maintenance of personal protective equipment when other controls are insufficient

Discussion questions:

1. Is personal protective equipment, appropriate to the hazard, provided at no cost to each worker exposed to dangerous chemicals?
2. How can workers be encouraged to properly select, use and maintain personal protective equipment?

Develop, periodically evaluate and carry out drills in emergency

Emergencies with chemicals can have catastrophic effects not only for workers but also for the community at large and the environment. Pre-emergency planning not only gives clear guidance to everyone in the enterprise as to what and what not to do, but it also provides an excellent opportunity for discussion with fire, police, medical and other emergency services outside the plant.

Management should establish emergency measures and facilities to deal with any eventuality. For example, in case of an accidental splash or contact with chemicals, emergency showers and eye-wash points should be provided in close proximity to the workstation (figure 56). These facilities should be regularly inspected to ensure that they are in operation when needed.

Similarly, in case of fire, suitable fire-fighting equipment should be provided to control the fire before fire-fighting units arrive. Workers who are involved in fire-fighting activities should receive adequate training. A fire evacuation plan should be established and rehearsed regularly to ensure smooth and speedy evacuation.

Management should also ensure that for every shift there are workers trained to administer first aid. They should check with the national legislation on the minimum number of first-aiders required for each shift. It should be noted that first aid can save lives or reduce the degree of injury if administered promptly.

Management has a responsibility to establish procedures to deal with emergencies and accidents that might arise from the use of hazardous chemicals at work. These procedures must be reviewed regularly and changed as appropriate when, for example:

Figure 56. Emergency showers and eye-wash points should be provided in close proximity to work-stations where hazardous chemicals are used

- new chemicals are brought into the workplace;
- new chemical processes are developed or existing processes are changed.

As already mentioned, there is also a need for management to establish a training programme for workers to deal with chemical emergencies.

A training programme may include the following:

- arrangements for raising the alarm;
- calling on the appropriate emergency assistance;
- use of appropriate personal protective equipment (and its limitations) to deal with the emergency;
- actions to evacuate anyone in immediate danger;
- the provision of life-saving first aid;
- the use of specialized equipment and materials including first-aid, fire-fighting, and spill and leak control equipment;
- actions to minimize the magnitude of the incident;
- actions to evacuate adjacent premises when necessary.

Discussion questions:

1. Outline your enterprise's plan for dealing with first-aid, fire or spill and leak situations.
2. Describe the equipment needed for dealing with chemical emergencies.

Establish and maintain exposure monitoring procedures, including medical surveillance

Management should establish a regular procedure to monitor the exposure levels of workers handling hazardous chemicals (figure 57). These levels should not exceed the acceptable limits promulgated by the national legislation. If exposure exceeds these limits, then immediate investigation should be conducted to determine the causes. Remedial actions should be taken promptly. During the period when corrective measures are being carried out, the workers must be provided with proper protective equipment or prevented from entering the contaminated area.

All exposure monitoring records should be kept and maintained in good order.

Workers exposed to chemicals should undergo medical surveillance, including periodic examinations, to monitor their health and determine whether working with chemicals has induced any adverse effects. Since most occupational diseases have a long latency period, surveillance provides an opportunity to detect these diseases at an early stage so that protective measures and appropriate treatment may be instituted. It is important to ensure that medical practitioners who administer these programmes have adequate training in occupational medicine.

All health records should be maintained in good order.

Discussion questions:

1. How does plant management provide for monitoring the working environment?
2. Describe the in-plant or inter-enterprise (shared) occupational health services.

Figure 57. Regular monitoring of the working environment allows for the identification of potentially harmful substances

Plan and implement training programmes

Training and education form an important element in the management of hazardous chemicals. The installation of safety devices or procedures and practices to control chemicals should be complemented with training and education if implementation of the chemical control programme is to be effective.

All those who work with toxic chemicals should be aware of the hazards, controls and procedures that apply. These include safe operating procedures, the use and care of personal protective equipment, and emergency and first-aid measures. Where a labelling system is used, the workers should be trained to interpret the information provided.

Training is particularly important for new workers. However, all workers should be retrained at regular intervals or whenever there is change in processes or procedures.

Discussion questions:

1. What kinds of training programme have been organized in your workplace during the past year to deal with safety and health in the use of chemicals?
2. Describe what strategies could be used so that chemical safety training programmes reach all workers.

Remember:

1. *Management has the responsibility to manage chemicals safely in its organization.*
2. *The general principles for the safe management of chemicals are:*
   - *know what chemicals are in use;*
   - *inform workers on the hazards of chemicals and precautionary measures to be taken;*
   - *design and adapt the workplace to the needs of workers.*
3. *The safe management of chemicals should include the following elements:*
   - *company goals;*
   - *a committee for the safe management of chemicals;*
   - *operational and organizational controls which include:*
     - *classification and labelling, handling, storage, transfer and disposal;*
     - *environmental and medical surveillance programmes;*
     - *training and education;*
     - *emergency measures.*
4. *The duties of workers are to cooperate with management in implementing the management programme, and to work safely and in such a way as not to endanger the safety of others.*

Discussion questions:

1. Describe the chemical safety management programme in your enterprise. List its elements.
2. How would you suggest improving the programme?

3. Describe the various means whereby workers and management can cooperate in ensuring successful implementation of a chemical safety management programme.

Suggested further reading

Chemical safety data sheets, produced by the ILO International Occupational Safety and Health Information Centre (CIS).

ILO. *Occupational exposure limits for airborne toxic substances*, Occupational Safety and Health Series, No. 37 (Geneva, 3rd ed., 1991).

—. *Occupational exposure to airborne substances harmful to health*, An ILO code of practice (Geneva, 1980).

Patty's industrial hygiene and toxicology, 4 vols. (New York, Wiley-Interscience, 3rd ed., 1991).

Annex 1

Training checklist for safe use of chemicals at work

How to use this checklist

1 If you are not looking at your own enterprise, you will need some general information. Ask the owner or manager any questions you have. You should learn about the main products and production methods, the number of workers (male and female), the hours of work (including lunch break, other breaks and overtime) and any important operational and labour problems.

2 Define the work areas to be checked. In the case of a small enterprise, the whole production area can be checked. In the case of a larger enterprise, particular work areas can be defined for separate checking.

3 Read through the checklist and spend some time walking around before starting to check.

4 Read each item carefully. Look for a way to apply the measure. If necessary, ask the owner or workers questions. If the measure has already been applied or it is not needed, mark *No* under "Do you propose action?". If you think the measure would be worth while, mark *Yes*. Use the space under *Remarks* to put a description of the measure or its location.

5 After you have finished, look again at the items you have marked *Yes*. Identify those items that require immediate or urgent attention. Mark *Priority* for these items.

6 Before finishing, make sure that for each item you have marked *No* or *Yes*, and that for some items marked *Yes* you have marked *Priority*.

Management of chemicals

1. Establish a clear organizational arrangement on safety and health in the use of chemicals that is made known to managers, workers and outsiders who have to deal with the enterprise.

Do you propose action?

☐ No ☐ Yes ☐ Priority

Remarks ______________________

____

2. Designate a person or a committee appointed by management to plan and coordinate activities on chemical safety.

Do you propose action?

☐ No ☐ Yes ☐ Priority

Remarks ______________________

____

3. Establish a procedure on the purchase of (both new and existing) chemicals for the enterprise.

Do you propose action?

☐ No ☐ Yes ☐ Priority

Remarks ______________________

____

4. Establish an inventory of chemicals which are used in the enterprise.

Do you propose action?

☐ No ☐ Yes ☐ Priority

Remarks ______________________

____

5. Make sure that the enterprise has in its possession chemical safety data sheets for all the chemicals used.

Do you propose action?

☐ No ☐ Yes ☐ Priority

Remarks ______________________

____

Hazard identification

6. Store flammable chemicals in such a way as to prevent the formation of flammable or explosive mixtures.

Do you propose action?

☐ No ☐ Yes ☐ Priority

Remarks ______________________

____

7. Eliminate any open flames within the area where flammable chemicals are used, transferred or stored.

Do you propose action?

☐ No ☐ Yes ☐ Priority

Remarks ______________________

____

8. Clean and maintain floor areas, workbenches and machinery surfaces free from oil deposits and dusts.

Do you propose action?

☐ No ☐ Yes ☐ Priority

Remarks ______________________

____

9. Ensure that passageways are well marked and clear of debris.

Do you propose action?

☐ No ☐ Yes ☐ Priority

Remarks ________________________________

10. Provide storage racks around workstations for raw materials and finished products.

Do you propose action?

☐ No ☐ Yes ☐ Priority

Remarks ________________________________

Operational control methods

11. Examine whether it is practicable to substitute less toxic chemicals for toxic ones.

Do you propose action?

☐ No ☐ Yes ☐ Priority

Remarks ________________________________

12. Ensure that processes emitting dusts, vapours or mists are enclosed.

Do you propose action?

☐ No ☐ Yes ☐ Priority

Remarks ________________________________

13. Ensure that workers are protected by isolating processes that emit dusts, vapours or mists from other areas of the factory.

Do you propose action?

☐ No ☐ Yes ☐ Priority

Remarks ________________________________

14. Ensure that local ventilation systems are installed and functioning to reduce contamination of the work area.

Do you propose action?

☐ No ☐ Yes ☐ Priority

Remarks ________________________________

15. Ensure that natural ventilation provides an adequate air exchange.

Do you propose action?

☐ No ☐ Yes ☐ Priority

Remarks ________________________________

16. Install fans and other mechanical devices to improve general ventilation.

Do you propose action?

☐ No ☐ Yes ☐ Priority

Remarks ________________________________

Personal protective equipment

17. Provide personal protective equipment when other operational control measures cannot eliminate the risk of workers being exposed to airborne contaminants.

Do you propose action?

☐ No ☐ Yes ☐ Priority

Remarks ________________________________

18. Provide the worker with appropriate eye and skin protection when there is a possibility of a chemical splash.

Do you propose action?

☐ No ☐ Yes ☐ Priority

Remarks ________________________________

19. Make organizational arrangements so that personal protective equipment is well maintained and inspected.

Do you propose action?

☐ No ☐ Yes ☐ Priority

Remarks ________________________________

Safe procedures and practices

20. Ensure that all chemicals are clearly labelled with the name of the supplier, the name and origin of the chemical, the danger symbol(s) and indications of danger, and phrases about the risk and safety advice relating to use.

Do you propose action?

☐ No ☐ Yes ☐ Priority

Remarks ________________________________

21. Relabel chemicals that have been dispensed into smaller containers.

Do you propose action?

☐ No ☐ Yes ☐ Priority

Remarks ______

22. Ensure that chemicals are stored in appropriate, sound containers.

Do you propose action?

☐ No ☐ Yes ☐ Priority

Remarks ______

23. Ensure that storage areas are well ventilated and are located away from sources of ignition.

Do you propose action?

☐ No ☐ Yes ☐ Priority

Remarks ______

24. Use appropriate devices to transport and transfer chemicals to ensure that hazards are not created during these processes.

Do you propose action?

☐ No ☐ Yes ☐ Priority

Remarks ______

25. Make sure small spills are immediately cleaned up and the area is safe to continue work.

Do you propose action?

☐ No ☐ Yes ☐ Priority

Remarks ______

26. Ensure that waste materials and empty containers previously used to store chemicals are safely disposed of so that they do not pose a risk to workers or the environment.

Do you propose action?

☐ No ☐ Yes ☐ Priority

Remarks ______

27. Ensure that procedures for the safe storage, transport and disposal of chemicals are made known by management to the workers concerned in writing.

Do you propose action?

☐ No ☐ Yes ☐ Priority

Remarks ______

Monitoring of exposure

28. Designate a person or arrange organizational measures for a programme to monitor workers' exposure at regular intervals.

Do you propose action?

☐ No ☐ Yes ☐ Priority

Remarks ______

Medical surveillance

29. Ensure that workers assigned for the first time to areas where hazardous chemicals are used are subjected to a pre-placement medical examination.

Do you propose action?

☐ No ☐ Yes ☐ Priority

Remarks ______

30. Ensure that workers concerned with handling designated kinds of hazardous chemicals are provided with regular medical examinations to monitor their health.

Do you propose action?

☐ No ☐ Yes ☐ Priority

Remarks ______

Training and education

31. Provide newly recruited workers handling chemicals with initial and refresher training on the hazards posed by the chemical, safe procedures and practices, and emergency procedures.

Do you propose action?

☐ No ☐ Yes ☐ Priority

Remarks ______

32. Ensure that labelling and instructions on chemical containers are in a written language understandable to workers.

Do you propose action?

☐ No ☐ Yes ☐ Priority

Remarks ______________________________

____

33. Provide training on the use, maintenance, cleaning and storage of personal protective equipment to the workers provided with this equipment.

Do you propose action?

☐ No ☐ Yes ☐ Priority

Remarks ______________________________

____

34. Ensure that essential training programmes regarding hazardous chemicals are repeated at regular intervals.

Do you propose action?

☐ No ☐ Yes ☐ Priority

Remarks ______________________________

____

Emergency measures

35. Provide adequate emergency equipment (eye-wash points and emergency showers) in good working order at strategic locations.

Do you propose action?

☐ No ☐ Yes ☐ Priority

Remarks ______________________________

____

36. Provide the appropriate type of fire extinguisher in an adequate number to control fire emergencies involving chemicals.

Do you propose action?

☐ No ☐ Yes ☐ Priority

Remarks ______________________________

____

37. Ensure that the company provides a team of trained personnel on every shift to extinguish small chemical fires.

Do you propose action?

☐ No ☐ Yes ☐ Priority

Remarks ______________________________

____

38. Formulate and make known to all workers the fire evacuation plan and hold regular drills.

Do you propose action?

☐ No ☐ Yes ☐ Priority

Remarks ______________________________

____

39. Make sure that the company provides a person trained to provide first aid on every shift.

Do you propose action?

☐ No ☐ Yes ☐ Priority

Remarks ______________________________

____

40. Provide well-equipped, clearly marked first-aid boxes and other appropriate first-aid equipment in sufficient quantity.

Do you propose action?

☐ No ☐ Yes ☐ Priority

Remarks ______________________________

____

Annex 2

The Chemicals Convention, 1990 (No. 170), and Recommendation, 1990 (No. 177)

Convention No. 170

Convention concerning safety in the use of chemicals at work

(extracts)

The General Conference of the International Labour Organization,

Having been convened at Geneva by the Governing Body of the International Labour Office, and having met in its 77th Session on 6 June 1990, and

Noting the relevant international labour Conventions and Recommendations and, in particular, the Benzene Convention and Recommendation, 1971, the Occupational Cancer Convention and Recommendation, 1974, the Working Environment (Air Pollution, Noise and Vibration) Convention and Recommendation, 1977, the Occupational Safety and Health Convention and Recommendation, 1981, the Occupational Health Services Convention and Recommendation, 1985, the Asbestos Convention and Recommendation, 1986, and the list of occupational diseases, as amended in 1980, appended to the Employment Injury Benefits Convention, 1964, and

Noting that the protection of workers from the harmful effects of chemicals also enhances the protection of the general public and the environment, and

Noting that workers have a need for, and right to, information about the chemicals they use at work, and

Considering that it is essential to prevent or reduce the incidence of chemically induced illnesses and injuries at work by:

(a) ensuring that all chemicals are evaluated to determine their hazards;

(b) providing employers with a mechanism to obtain from suppliers information about the chemicals used at work so that they can implement effective programmes to protect workers from chemical hazards;

(c) providing workers with information about the chemicals at their workplaces, and about appropriate preventive measures so that they can effectively participate in protective programmes;

(d) establishing principles for such programmes to ensure that chemicals are used safely, and

Having regard to the need for cooperation within the International Programme on Chemical Safety between the International Labour Organization, the United Nations Environment Programme and the World Health Organization as well as with the Food and Agriculture Organization of the United Nations and the United Nations Industrial Development Organization, and noting the relevant instruments, codes and guidelines promulgated by these organizations, and

Having decided upon the adoption of certain proposals with regard to safety in the use of chemicals at work, which is the fifth item on the agenda of the session, and

Having determined that these proposals shall take the form of an international Convention;

adopts this twenty-fifth day of June of the year one thousand nine hundred and ninety the following Convention, which may be cited as the Chemicals Convention, 1990.

Part I. Scope and definitions

Article 1

1. This Convention applies to all branches of economic activity in which chemicals are used.

2. The competent authority of a Member ratifying this Convention, after consulting the most representative organizations of employers and workers concerned, and on the basis of an assessment of the hazards involved and the protective measures to be applied:

(a) may exclude particular branches of economic activity, undertaking or products from the application of the Convention, or certain provisions thereof, when:

  (i) special problems of a substantial nature arise; and

  (ii) the overall protection afforded in pursuance of national law and practice is not inferior to that which would result from the full application of the provisions of the Convention;

(b) shall make special provision to protect confidential information whose disclosure to a competitor would be liable to cause harm to an employer's business so long as the safety and health of workers are not compromised thereby.

3. This Convention does not apply to articles which will not expose workers to a hazardous chemical under normal or reasonably foreseeable conditions of use.

4. This Convention does not apply to organisms, but does apply to chemicals derived from organisms.

Article 2

For the purposes of this Convention:

(a) the term "chemicals" means chemical elements and compounds, and mixtures thereof, whether natural or synthetic;

(b) the term "hazardous chemical" includes any chemical which has been classified as hazardous in accordance with Article 6 or for which relevant information exists to indicate that the chemical is hazardous;

(c) the term "use of chemicals at work" means any work activity which may expose a worker to a chemical, including:

  (i) the production of chemicals;

(ii) the handling of chemicals;
(iii) the storage of chemicals;
(iv) the transport of chemicals;
(v) the disposal and treatment of waste chemicals;
(vi) the release of chemicals resulting from work activities;
(vii) the maintenance, repair and cleaning of equipment and containers for chemicals;

(d) the term "branches of economic activity" means all branches in which workers are employed, including the public service;

(e) the term "article" means an object which is formed to a specific shape or design during its manufacture or which is in its natural shape, and whose use in that form is dependent in whole or in part on its shape or design;

(f) the term "workers' representatives" means persons who are recognized as such by national law or practice, in accordance with the Workers' Representatives Convention, 1971.

Part II. General principles

Article 3

The most representative organizations of employers and workers concerned shall be consulted on the measures to be taken to give effect to the provisions of this Convention.

Article 4

In the light of national conditions and practice and in consultation with the most representative organizations of employers and workers, each Member shall formulate, implement and periodically review a coherent policy on safety in the use of chemicals at work.

Article 5

The competent authority shall have the power, if justified on safety and health grounds, to prohibit or restrict the use of certain hazardous chemicals, or to require advance notification and authorization before such chemicals are used.

Part III. Classification and related measures

Article 6

Classification systems

1. Systems and specific criteria appropriate for the classification of all chemicals according to the type and degree of their intrinsic health and physical hazards and for assessing the relevance of the information required to determine whether a chemical is hazardous shall be established by the competent authority, or by a body approved or recognized by the competent authority, in accordance with national or international standards.

2. The hazardous properties of mixtures composed of two or more chemicals may be determined by assessments based on the intrinsic hazards of their component chemicals.

3. In the case of transport, such systems and criteria shall take into account the United Nations Recommendations on the transport of dangerous goods.

4. The classification systems and their application shall be progressively extended.

Article 7

Labelling and marking

1. All chemicals shall be marked so as to indicate their identity.

2. Hazardous chemicals shall in addition be labelled, in a way easily understandable to the workers, so as to provide essential information regarding their classification, the hazards they present and the safety precautions to be observed.

3. (1) Requirements for marking or labelling chemicals pursuant to paragraphs 1 and 2 of this Article shall be established by the competent authority, or by a body approved or recognized by the competent authority, in accordance with national or international standards.

(2) In the case of transport, such requirements shall take into account the United Nations Recommendations on the transport of dangerous goods.

Article 8

Chemical safety data sheets

1. For hazardous chemicals, chemical safety data sheets containing detailed essential information regarding their identity, supplier, classification, hazards, safety precautions and emergency procedures shall be provided to employers.

2. Criteria for the preparation of chemical safety data sheets shall be established by the competent authority, or by a body approved or recognized by the competent authority, in accordance with national or international standards.

3. The chemical or common name used to identify the chemical on the chemical safety data sheet shall be the same as that used on the label.

Article 9

Responsibilities of suppliers

1. Suppliers of chemicals, whether manufacturers, importers or distributors, shall ensure that:

(a) such chemicals have been classified in accordance with Article 6 on the basis of knowledge of their properties and a search of available information or assessed in accordance with paragraph 3 below;

(b) such chemicals are marked so as to indicate their identity in accordance with Article 7, paragraph 1;

(c) hazardous chemicals they supply are labelled in accordance with Article 7, paragraph 2;

(d) chemical safety data sheets are prepared for such hazardous chemicals in accordance with Article 8, paragraph 1, and provided to employers.

2. Suppliers of hazardous chemicals shall ensure that revised labels and chemical safety data sheets are prepared and provided to employers, by a method

which accords with national law and practice, whenever new relevant safety and health information becomes available.

3. Suppliers of chemicals which have not yet been classified in accordance with Article 6 shall identify the chemicals they supply and assess the properties of these chemicals on the basis of a search of available information in order to determine whether they are hazardous chemicals.

Part IV. Responsibilities of employers

Article 10

Identification

1. Employers shall ensure that all chemicals used at work are labelled or marked as required by Article 7 and that chemical safety data sheets have been provided as required by Article 8 and are made available to workers and their representatives.

2. Employers receiving chemicals that have not been labelled or marked as required under Article 7, or for which chemical safety data sheets have not been provided as required under Article 8, shall obtain the relevant information from the supplier or from other reasonably available sources, and shall not use the chemicals until such information is obtained.

3. Employers shall ensure that only chemicals which are classified in accordance with Article 6 or identified and assessed in accordance with Article 9, paragraph 3, and labelled or marked in accordance with Article 7 are used and that any necessary precautions are taken when they are used.

4. Employers shall maintain a record of hazardous chemicals used at the workplace, cross-referenced to the appropriate chemical safety data sheets. This record shall be accessible to all workers concerned and their representatives.

Article 11

Transfer of chemicals

Employers shall ensure that when chemicals are transferred into other containers or equipment, the contents are indicated in a manner which will make known to workers their identity, any hazards associated with their use and any safety precautions to be observed.

Article 12

Exposure

Employers shall:

(a) ensure that workers are not exposed to chemicals to an extent which exceeds exposure limits or other exposure criteria for the evaluation and control of the working environment established by the competent authority, or by a body approved or recognized by the competent authority, in accordance with national or international standards;

(b) assess the exposure of workers to hazardous chemicals;

(c) monitor and record the exposure of workers to hazardous chemicals when this is necessary to safeguard their safety and health or as may be prescribed by the competent authority;

(d) ensure that the records of the monitoring of the working environment and of the exposure of workers using hazardous chemicals are kept for a period prescribed by the competent authority and are accessible to the workers and their representatives.

Article 13

Operational control

1. Employers shall make an assessment of the risks arising from the use of chemicals at work, and shall protect workers against such risks by appropriate means, such as:

(a) the choice of chemicals that eliminate or minimize the risk:

(b) the choice of technology that eliminates or minimizes the risk;

(c) the use of adequate engineering control measures;

(d) the adoption of working systems and practices that eliminate or minimize the risk;

(e) the adoption of adequate occupational hygiene measures;

(f) where recourse to the above measures does not suffice, the provision and proper maintenance of personal protective equipment and clothing at no cost to the worker, and the implementation of measures to ensure their use.

2. Employers shall:

(a) limit exposure to hazardous chemicals so as to protect the safety and health of workers;

(b) provide first aid;

(c) make arrangements to deal with emergencies.

Article 14

Disposal

Hazardous chemicals which are no longer required and containers which have been emptied but which may contain residues of hazardous chemicals shall be handled or disposed of in a manner which eliminates or minimizes the risk to safety and health and to the environment, in accordance with national law and practice.

Article 15

Information and training

Employers shall:

(a) inform the workers of the hazards associated with exposure to chemicals used at the workplace;

(b) instruct the workers how to obtain and use the information provided on labels and chemical safety data sheets;

(c) use the chemical safety data sheets, along with information specific to the workplace, as a basis for the preparation of instructions to workers, which should be written if appropriate;

(d) train the workers on a continuing basis in the practices and procedures to be followed for safety in the use of chemicals at work.

Article 16

Cooperation

Employers, in discharging their responsibilities, shall cooperate as closely as possible with workers or their representatives with respect to safety in the use of chemicals at work.

Part V. Duties of workers

Article 17

1. Workers shall cooperate as closely as possible with their employers in the discharge by the employers of their responsibilities and comply with all procedures and practices relating to safety in the use of chemicals at work.

2. Workers shall take all reasonable steps to eliminate or minimize risk to themselves and to others from the use of chemicals at work.

Part VI. Rights of workers and their representatives

Article 18

1. Workers shall have the right to remove themselves from danger resulting from the use of chemicals when they have reasonable justification to believe there is an imminent and serious risk to their safety or health, and shall inform their supervisor immediately.

2. Workers who remove themselves from danger in accordance with the provisions of the previous paragraph or who exercise any other rights under this Convention shall be protected against undue consequences.

3. Workers concerned and their representatives shall have the right to:
(a) information on the identity of chemicals used at work, the hazardous properties of such chemicals, precautionary measures, education and training;
(b) the information contained in labels and markings;
(c) chemical safety data sheets;
(d) any other information required to be kept by this Convention.

4. Where disclosure of the specific identity of an ingredient of a chemical mixture to a competitor would be liable to cause harm to the employer's business, the employer may, in providing the information required under paragraph 3 above, protect that identity in a manner approved by the competent authority under Article 1, paragraph 2 (b).

Part VII. Responsibility of exporting States

Article 19

When in an exporting member State all or some uses of hazardous chemicals are prohibited for reasons of safety and health at work, this fact and the reasons for it shall be communicated by the exporting member State to any importing country.

. . .

Recommendation No. 177

Recommendation concerning safety in the use of chemicals at work

The General Conference of the International Labour Organization,

Having been convened at Geneva by the Governing Body of the International Labour Office, and having met in its 77th Session on 6 June 1990, and

Having decided upon the adoption of certain proposals with regard to safety in the use of chemicals at work, which is the fifth item on the agenda of the session, and

Having determined that these proposals shall take the form of a Recommendation supplementing the Chemicals Convention, 1990;

adopts this twenty-fifth day of June of the year one thousand nine hundred and ninety the following Recommendation, which may be cited as the Chemicals Recommendation, 1990.

I. General provisions

1. The provisions of this Recommendation should be applied in conjunction with those of the Chemicals Convention, 1990 (hereafter referred to as "the Convention").

2. The most representative organizations of employers and workers concerned should be consulted on the measures to be taken to give effect to the provisions of this Recommendation.

3. The competent authority should specify categories of workers who for reasons of safety and health are not allowed to use specified chemicals or are allowed to use them only under conditions prescribed in accordance with national laws or regulations.

4. The provisions of this Recommendation should also apply to such self-employed persons as may be specified by national laws or regulations.

5. The special provisions established by the competent authority to protect confidential information, under Article 1, paragraph 2 (b), and Article 18, paragraph 4, of the Convention, should:
(a) limit the disclosure of confidential information to those who have a need related to workers' safety and health;
(b) ensure that those who obtain confidential information agree to use it only to address safety and health needs and otherwise to protect its confidentiality;

(c) provide that relevant confidential information be disclosed immediately in an emergency;
(d) provide for procedures to consider promptly the validity of the confidentiality claim and of the need for the information withheld where there is a disagreement regarding disclosure.

II. Classification and related measures

Classification

6. The criteria for the classification of chemicals established pursuant to Article 6, paragraph 1, of the Convention should be based upon the characteristics of chemicals including:
(a) toxic properties, including both acute and chronic health effects in all parts of the body;
(b) chemical or physical characteristics, including flammable, explosive, oxidizing and dangerously reactive properties;
(c) corrosive and irritant properties;
(d) allergenic and sensitizing effects;
(e) carcinogenic effects;
(f) teratogenic and mutagenic effects;
(g) effects on the reproductive system.

7. (1) As far as is reasonably practicable, the competent authority should compile and periodically update a consolidated list of the chemical elements and compounds used at work, together with relevant hazard information.

(2) For chemical elements and compounds not yet included in the consolidated list, the manufacturers or importers should, unless exempted, be required to transmit to the competent authority, prior to use at work, and in a manner consistent with the protection of confidential information under Article 1, paragraph 2 (b), of the Convention, such information as is necessary for the maintenance of the list.

Labelling and marking

8. (1) The requirements for the labelling and marking of chemicals established pursuant to Article 7 of the Convention should be such as to enable persons handling or using chemicals to recognize and distinguish between them both when receiving and when using them, so that they may be used safely.

(2) The labelling requirements for hazardous chemicals should, in conformity with existing national or international systems, cover:
(a) the information to be given on the label including as appropriate:
 (i) trade names;
 (ii) identity of the chemical;
 (iii) name, address and telephone number of the supplier;
 (iv) hazard symbols;
 (v) nature of the special risks associated with the use of the chemical;
 (vi) safety precautions;
 (vii) identification of the batch;
 (viii) the statement that a chemical safety data sheet giving additional information is avilable from the employer;
 (ix) the classification assigned under the system established by the competent authority;
(b) the legibility, durability and size of the label;
(c) the uniformity of labels and symbols, including colours.

(3) The label should be easily understandable by workers.

(4) In the case of chemicals not covered by subparagraph (2) above, the marking may be limited to the identity of the chemical.

9. Where it is impracticable to label or mark a chemical in view of the size of the container or the nature of the package, provision should be made for other effective means of recognition such as tagging or accompanying documents. However, all containers of hazardous chemicals should indicate the hazards of the contents through appropriate wording or symbols.

Chemical safety data sheets

10. (1) The criteria for the preparation of chemical safety data sheets for hazardous chemicals should ensure that they contain essential information including, as applicable:
(a) chemical product and company identification (including trade or common name of the chemical and details of the supplier or manufacturer);
(b) composition/information on ingredients (in a way that clearly identifies them for the purpose of conducting a hazard evaluation);
(c) hazards identification;
(d) first-aid measures;
(e) firefighting measures;
(f) accidental release measures;
(g) handling and storage;
(h) exposure controls/personal protection (including possible methods of monitoring workplace exposure);
(i) physical and chemical properties;
(j) stability and reactivity;
(k) toxicological information (including the potential routes of entry into the body and the possibility of synergism with other chemicals or hazards encountered at work);
(l) ecological information;
(m) disposal considerations;
(n) transport information;
(o) regulatory information;
(p) other information (including the date of preparation of the chemical safety data sheet).

(2) Where the names or concentrations of the ingredients referred to in subparagraph (1) (b) above constitute confidential information, they may, in accordance with Article 1, paragraph 2 (b), of the Convention, be omitted from the chemical safety data sheet. In accordance with Paragraph 5 of this Recommendation the information should be disclosed

on request and in writing to the competent authority and to concerned employers, workers and their representatives who agree to use the information only for the protection of workers' safety and health and not otherwise to disclose it.

III. Responsibilities of employers

Monitoring of exposure

11. (1) Where workers are exposed to hazardous chemicals, the employer should be required to:

(a) limit exposure to such chemicals so as to protect the health of workers;

(b) assess, monitor and record, as necessary, the concentration of airborne chemicals at the workplace.

(2) Workers and their representatives and the competent authority should have access to these records.

(3) Employers should keep the records provided for in this Paragraph for a period of time determined by the competent authority.

Operational control within the workplace

12. (1) Measures should be taken by employers to protect workers against hazards arising from the use of chemicals at work, based upon the criteria established pursuant to Paragraphs 13 to 16 below.

(2) In accordance with the Tripartite Declaration of Principles concerning Multinational Enterprises and Social Policy, adopted by the Governing Body of the International Labour Office, a national or multinational enterprise with more than one establishment should provide safety measures relating to the prevention and control of, and protection against, health hazards due to occupational exposure to hazardous chemicals, without discrimination, to the workers in all its establishments regardless of the place or country in which they are situated.

13. The competent authority should ensure that criteria are established for safety in the use of hazardous chemicals, including provisions covering, as applicable:

(a) the risk of acute or chronic diseases due to entry into the body by inhalation, skin absorption or ingestion;

(b) the risk of injury or disease from skin or eye contact;

(c) the risk of injury from fire, explosion or other events resulting from physical properties or chemical reactivity;

(d) the precautionary measures to be taken through:

(i) the choice of chemicals that eliminate or minimize such risks;

(ii) the choice of processes, technology and installations that eliminate or minimize such risks;

(iii) the use and proper maintenance of engineering control measures;

(iv) the adoption of working systems and practices that eliminate or minimize such risks;

(v) the adoption of adequate personal hygiene measures and provision of adequate sanitary facilities;

(vi) the provision, maintenance and use of suitable personal protective equipment and clothing, at no cost to the worker where the above measures have not proved sufficient to eliminate such risks;

(vii) the use of signs and notices;

(viii) adequate preparations for emergencies.

14. The competent authority should ensure that criteria are established for safety in the storage of hazardous chemicals, including provisions covering, as applicable:

(a) the compatibility and segregation of stored chemicals;

(b) the properties and quantity of chemicals to be stored;

(c) the security and siting of and access to stores;

(d) the construction, nature and integrity of storage containers;

(e) loading and unloading of storage containers;

(f) labelling and relabelling requirements;

(g) precautions against accidental release, fire, explosion and chemical reactivity;

(h) temperature, humidity and ventilation;

(i) precautions and procedures in case of spillage;

(j) emergency procedures;

(k) possible physical and chemical changes in stored chemicals.

15. The competent authority should ensure that criteria consistent with national or international transport regulations are established for the safety of workers involved in the transport of hazardous chemicals, including provisions covering, as applicable:

(a) the properties and quantity of chemicals to be transported;

(b) the nature, integrity and protection of packagings and containers used in transport, including pipelines;

(c) the specifications of the vehicle used in transport;

(d) the routes to be taken;

(e) the training and qualifications of transport workers;

(f) labelling requirements;

(g) loading and unloading;

(h) procedures in case of spillage.

16. (1) The competent authority should ensure that criteria consistent with national or international regulations regarding disposal of hazardous waste are established for procedures to be followed in the disposal and treatment of hazardous chemicals and hazardous waste products with a view to ensuring the safety of workers.

(2) These criteria should include provisions covering, as applicable:

(a) the method of identification of waste products;

(b) the handling of contaminated containers;

(c) the identification, construction, nature, integrity and protection of waste containers;

(d) the effects on the working environment;
(e) the demarcation of disposal areas;
(f) the provision, maintenance and use of personal protective equipment and clothing;
(g) the method of disposal or treatment.

17. The criteria for the use of chemicals at work established pursuant to the provisions of the Convention and this Recommendation should be as consistent as possible with the protection of the general public and the environment and any criteria established for that purpose.

Medical surveillance

18. (1) The employer, or the institution competent under national law and practice, should be required to arrange, through a method which accords with national law and practice, such medical surveillance of workers as is necessary:
(a) for the assessment of the health of workers in relation to hazards caused by exposure to chemicals;
(b) for the diagnosis of work-related diseases and injuries caused by exposure to hazardous chemicals.

(2) Where the results of medical tests or investigations reveal clinical or preclinical effects, measures should be taken to prevent or reduce exposure of the workers concerned, and to prevent further deterioration of their health.

(3) The results of medical examinations should be used to determine health status with respect to exposure to chemicals, and should not be used to discriminate against the worker.

(4) Records resulting from medical surveillance of workers should be kept for a period of time and by persons prescribed by the competent authority.

(5) Workers should have access to their own medical records, either personally or through their own physicians.

(6) The confidentiality of individual medical records should be respected in accordance with generally accepted principles of medical ethics.

(7) The results of medical examinations should be clearly explained to the workers concerned.

(8) Workers and their representatives should have access to the results of studies prepared from medical records, where individual workers cannot be identified.

(9) The results of medical records should be made available to prepare appropriate health statistics and epidemiological studies, provided anonymity is maintained, where this may aid in the recognition and control of occupational diseases.

First aid and emergencies

19. In accordance with any requirements laid down by the competent authority, employers should be required to maintain procedures, including first-aid arrangements, to deal with emergencies and accidents resulting from the use of hazardous chemicals at work and to ensure that workers are trained in these procedures.

IV. Cooperation

20. Employers, workers and their representatives should cooperate as closely as possible in the application of measures prescribed pursuant to this Recommendation.

21. Workers should be required to:
(a) take care as far as possible of their own safety and health and of that of other persons who may be affected by their acts or omissions at work in accordance with their training and with instructions given by their employer;
(b) use properly all devices provided for their protection or the protection of others;
(c) report forthwith to their supervisor any situation which they believe could present a risk, and which they cannot deal with themselves.

22. Publicity material concerning hazardous chemicals intended for use at work should call attention to their hazards and the necessity to take precautions.

23. Suppliers should, on request, provide employers with such information as is available and required for the evaluaton of any unusual hazards which might result from a particular use of a chemical at work.

V. Rights of workers

24. (1) Workers and their representatives should have the right to:
(a) obtain chemical safety data sheets and other information from the employer so as to enable them to take adequate precautions, in cooperation with their employer, to protect workers against risks from the use of hazardous chemicals at work;
(b) request and participate in an investigation by the employer or the competent authority of possible risks resulting from the use of chemicals at work.

(2) Where the information requested is confidential in accordance with Article 1, paragraph 2 (b), and Article 18, paragraph 4, of the Convention, employers may require the workers or workers' representatives to limit its use to the evaluation and control of possible risks arising from the use of chemicals at work, and to take reasonable steps to ensure that this information is not disclosed to potential competitors.

(3) Having regard to the Tripartite Declaration of Principles concerning Multinational Enterprises and Social Policy, multinational enterprises should make available, upon request, to workers concerned, workers' representatives, the competent authority and employers' and workers' organizations in all countries in which they operate, information on the standards and procedures related to the use of hazardous chemicals, relevant to their local operations, which they observe in other countries.

25. (1) Workers should have the right:
(a) to bring to the attention of their representatives, the employer or the competent authority, potential hazards arising from the use of chemicals at work;
(b) to remove themselves from danger resulting from the use of chemicals when they have reasonable justification to believe there is an imminent and

serious risk to their safety or health, and should inform their supervisor immediately;

(c) in the case of a health condition, such as chemical sensitization, placing them at increased risk of harm from a hazardous chemical, to alternative work not involving that chemical, if such work is available and if the workers concerned have the qualifications or can reasonably be trained for such alternative work;

(d) to compensation if the case referred to in subparagraph (1) (c) results in loss of employment;

(e) to adequate medical treatment and compensation for injuries and diseases resulting from the use of chemicals at work.

(2) Workers who remove themselves from danger in accordance with the provisions of subparagraph (1) (b) or who exercise any of their rights under this Recommendation should be protected against undue consequences.

(3) Where workers have removed themselves from danger in accordance with subparagraph (1) (b), the employer, in cooperation with workers and their representatives, should immediately investigate the risk and take any corrective steps necessary.

(4) Women workers should have the right, in the case of pregnancy or lactation, to alternative work not involving the use of, or exposure to, chemicals hazardous to the health of the unborn or nursing child, where such work is available, and the right to return to their previous jobs at the appropriate time.

26. Workers should receive:

(a) information on the classification and labelling of chemicals and on chemical safety data sheets in forms and languages which they easily understand;

(b) information on the risks which may arise from the use of hazardous chemicals in the course of their work;

(c) instruction, written or oral, based on the chemical safety data sheet and specific to the workplace if appropriate;

(d) training and, where necessary, retraining in the methods which are available for the prevention and control of, and for protection against, such risks, including correct methods of storage, transport and waste disposal as well as emergency and first-aid measures.

Annex 3

Chemical safety data sheets

The following pages give examples of chemical safety data sheets published by the International Occupational Safety and Health Information Centre (CIS) of the ILO. These sheets may be obtained from:

ILO-CIS
CH-1211 Geneva 22
Tel. + 41 22 799 67 40
Telex 415 647 ILO CH
Telefax + 41 22 798 62 53

CHEMICAL INFO-SHEET

CS-1 **BENZENE**

CAS 71-43-2
FORMULA: C_6-H_6

DESCRIPTION

Colourless liquid with sweet odour.
Used to produce:
- dyes
- plastics
- textiles
- detergents
- paints
- other chemicals

Used as a solvent for paints and adhesives.
Present in small amounts in gasoline.
Industrial uses are decreasing.

SHORT-TERM EXPOSURE EFFECTS

Very Toxic

Inhalation:
A 5-hour exposure at 50-150 ppm can cause:
- headache
- tiredness

A 1-hour exposure at 200-500 ppm can cause:
- nausea
- dizziness
- confusion

A 30-60 minute exposure at 3000 ppm can cause nose and throat irritation.
A 30-minute exposure at 7500 ppm can cause death.

Eye Contact:
High concentrations of vapour cause slight irritation.
Liquid causes a slight burning sensation.

Skin Contact:
Liquid dissolves skin oils and causes irritation and blistering.

Ingestion:
May cause the same symptoms as inhalation.
If swallowed, liquid drawn into lungs can cause severe injury.

LONG-TERM EXPOSURE EFFECTS

Benzene can damage the blood-forming system causing:
- anemia
- infections
- bruising
- bleeding

Prolonged low-level exposure can cause:
- hearing damage
- headache
- dizziness
- tiredness
- paleness
- problems with vision and balance

Repeated skin contact causes:
- redness
- drying
- blistering

Known to cause cancer in humans.
Cancers of the white-blood cells can develop.
Reproductive effects such as menstrual problems may result.
Genetic damage can develop after long-term, severe exposures.

FIRE AND EXPLOSION

Flammable Liquid

Highly flammable.
Dangerous fire hazard.
Extinguish fires with:
- dry chemical
- foam
- carbon dioxide

Vapours can travel at ground level to ignition source and flash back.

CHEMICAL REACTIVITY

Normally stable.
Contact with strong oxidizers, such as nitric acid, increases risk of fire and explosion.

PERSONAL PROTECTION

Inhalation:
Wear a self-contained breathing apparatus or a supplied-air respirator if vapour or mist concentration is unknown or present at any detectable concentration.

Skin:
Wear, as needed:
- gloves
- coveralls
- boots

A suitable material is Viton.
Have a safety shower/eye-wash fountain available in the immediate area.

Eyes:
Wear chemical safety goggles.
A face shield may also be necessary.

STORAGE AND HANDLING

Follow rules for storing and handling flammable liquids.
Store benzene:
- in tightly-closed, grounded, labelled containers
- in a cool, dry, well-ventilated area
- out of direct sunlight
- away from incompatible materials and heat.

Use non-sparking ventilation systems and electrical equipment.
Use in small quantities in designated areas.
Prevent release of vapours into workplace air.

CLEAN-UP AND DISPOSAL

Only trained personnel should clean up.
Ensure appropriate ventilation is provided.
Use appropriate protective clothing and respirators.
Stop or reduce leak if possible.
Absorb small spills with sand or other inert material.
Place in suitable, covered containers.
Flush area with water.
For large spills, contact emergency services and supplier for advice.
Comply with environmental regulations.

FIRST AID

Inhalation:
Remove source of benzene or move victim to fresh air.
If breathing has stopped, begin artificial respiration.

Eye Contact:
Flush affected eye with lukewarm, gently flowing water for 20 minutes, holding the eyelid open.
Do not rinse contaminated water into non-affected eye.

Skin Contact:
Remove contaminated clothing.
Gently blot or brush away excess chemical quickly.
Wash gently and thoroughly with water and non-abrasive soap.

Ingestion:
Never give anything by mouth if victim is:
- losing consciousness
- unconscious
- convulsing

Rinse mouth thoroughly with water.
Have victim drink about 250 mL (8 oz.) of water.
DO NOT INDUCE VOMITING.
If vomiting occurs, have victim lean forward and repeat administration of water.

Note: *Obtain medical attention IMMEDIATELY for all serious exposures. Consult a physician or the nearest Poison Control Centre.*

NEED MORE INFORMATION?

See CHEMINFO record no. 179E, Chemical Hazard Summary No. 34, available from CCOHS.

This document was originally published by CCOHS (Canadian Centre for Occupational Health and Safety) in its **Chemical Infogram series**
Further information can be obtained from CIS or its national centres.

CHEMICAL INFO-SHEET

CS-21 **CHLORINE**

CAS 7782-50-5
FORMULA: Cl-Cl

DESCRIPTION

Greenish-yellow gas or amber liquid (under pressure).
Pungent odour.
Used in producing:
- chlorinated chemicals
- pesticides
- refrigerants
- plastics
- bleach

Used in:
- water purification
- sewage disinfection
- food processing

SHORT-TERM EXPOSURE EFFECTS

Inhalation:
Causes severe nose, throat and upper respiratory tract irritation.
Symptoms include:
- itchy nose (0.2 ppm)
- dry throat, coughing and difficulty breathing (1.0 ppm)
- shortness of breath, headache (above 1.3 ppm)
- intense choking, chest pain and vomiting (above 30 ppm)

Severe exposure causes:
- bronchitis
- fluid in the lungs
- death (above 1000 ppm)

Eye Contact:
Severe eye irritant.
Gas causes:
- stinging
- burning sensation with tearing

Liquid can cause:
- burns
- permanent damage
- possibly blindness

Skin Contact:
Severe skin irritant.
High gas concentrations cause:
- burning
- reddening
- blisters

Liquid causes:
- burns
- possibly frostbite

Ingestion:
Liquid may cause:
- pain
- burning
- thirst
- abdominal cramps
- nausea

LONG-TERM EXPOSURE EFFECTS

May cause:
- respiratory effects
- irritation of the nose
- corrosion of tooth enamel

FIRE AND EXPLOSION

Chlorine can support combustion and is a serious fire risk.
Extinguish fires with:
- dry chemical
- carbon dioxide

Chlorine gas will collect in low-lying areas.

CHEMICAL REACTIVITY

Extremely reactive.
Reacts violently with:
- many combustible materials
- other chemicals including water

Reacts vigorously with:
- hydrocarbons
- some finely powdered metals
- nitrogen compounds

Corrosive to most metals in the presence of water.

PERSONAL PROTECTION

Inhalation:
Wear suitable respirator if gas concentration is unknown or exceeds exposure limits.

Skin:
Wear:
- gloves
- coveralls
- boots

Suitable materials may be:
- Viton
- polyvinyl chloride (PVC)

Have a safety shower/eyewash fountain available in the immediate area.

Eyes:
Wear non-ventilated chemical-splash goggles.
A full face shield may also be necessary.

STORAGE AND HANDLING

Follow rules for storing and handling compressed gases and oxidizing materials
Store chlorine:
- in labelled, steel, pressure cylinders
- secured in an upright position
- in a cool (below 50°C), dry area away from combustibles, ignition sources and incompatible materials

Handle cylinders carefully according to manufacturer's recommendations.

CLEAN-UP AND DISPOSAL

Only trained personnel should clean up.
Ensure appropriate ventilation is provided.
Use appropriate protective clothing and respirators.
Follow manufacturer's recommendations for clean-up and neutralization.
For disposal, comply with environmental regulations.

FIRST AID

Inhalation:
Ensure your own safety before attempting rescue.
Remove source of chlorine or move victim to fresh air.
If breathing has stopped begin artificial respiration immediately.
Avoid mouth-to-mouth contact.
If heart has stopped begin cardio-pulmonary resuscitation (CPR) immediately.
Trained person may administer oxygen if physician advises.

Eye Contact:
Flush affected eye with lukewarm, gently flowing water for 30 minutes, holding the eyelid open.
Do not rinse contaminated water into non-affected eye.

Skin Contact:
Avoid contact.
Wear impervious protective clothing.
Flush affected area with lukewarm, gently running water for at least 20 minutes.
Under running water, remove contaminated clothing.

Ingestion:
Never give anything by mouth if victim is:
- losing consciousness
- unconscious
- convulsing

Rinse mouth thoroughly with water.
Have victim drink about 250 mL (8 oz.) of water.
DO NOT INDUCE VOMITING.
If vomiting occurs:
- rinse mouth
- repeat administration of water

Note: *Obtain medical attention IMMEDIATELY for all serious exposures. Consult a physician or the nearest Poison Control Centre.*

NEED MORE INFORMATION?

See CHEMINFO record no. 85E, available from CCOHS.

This document was originally published by CCOHS (Canadian Centre for Occupational Health and Safety) in its **Chemical Infogram series**
Further information can be obtained from CIS or its national centres.

Annex 4

Classification, identification and labelling of chemicals

J. Takala, Head of the International Occupational Safety and Health Information Centre (CIS), ILO

This document is intended for guidance and is not intended to replace national regulations. It describes one system of classification and labelling, that of the European Communities (EC).

Labelling of dangerous substances

1. General information

In this guide a safety label is assumed to be attached to the package of a substance hazardous to health. This label contains information on the dangerous nature of the substance and instructions for handling it safely.

The marking of the package should be prescribed in national regulations concerning the identification and labelling system of substances hazardous to health.

The safety label is used to mark the substances and compounds as presented on the lists included in the national regulations. On the list, danger symbols, risk phrases and safety advice phrases are given for each one of the substances (those of the EC are stated in this standard).[1]

2. Danger symbols[2]

2.1. Significance

In the lists of the national regulations, the letter symbols have the following significance, and the danger symbols shown opposite are presented as follows:

Letter symbol	*Significance*
E	Explosive
O	Oxidizing
F	Highly flammable
T	Toxic
C	Corrosive
X_n	Harmful (less than T)
X_i	Irritant (less than C)

2.2. Colour

The pictorial symbol indicating danger is black, the background orange.

2.3. Size

The size of the danger symbol should cover at least one-tenth of the surface of the safety label. When two danger symbols are used, their total size should also be at least one-tenth of the label. The minimum size of the danger symbol should, however, be 1 cm^2 irrespective of the number of danger symbols.

3. Safety label

The safety label can be separated from the rest of the cover by means of black lining. The danger markings (e.g. danger symbols, risk phrase and safety advice phrases) should be clearly visible against the background. The marking may be done by means of a printed label or on the package itself.

3.1. Content of the label

The safety label should give information on:

- dangerous substances and their concentration;
- danger marking, e.g. danger symbols, risk phrases (R-phrases) and safety advice phrase (S-phrases) – see below;
- the manner of destroying the package and, if necessary, instructions for rendering it innocuous;
- the trade name of the substance.

The safety label should also contain:

- the manufacturer, packer or importer of the substance with their respective addresses;
- the quantity of the substance in the package;

unless this information is stated elsewhere on the cover of the package.

3.2. Danger marking

The markings of the substances itemized in the national regulations can be obtained from the list enclosed therein. The compounds and products on the list containing the respective substances should be marked in case they have effects corresponding to that of some substance mentioned in the list. In case the product is a compound containing an amount of a dangerous substance that is below the limit to require the danger symbol T or C, the symbol X_n or X_i can be put on the product as an indication of a lesser danger. The danger markings are built up by collecting the markings for different substances from the list. If the compound or the product is then liable to several danger symbols, not more than the two symbols indicating the greatest danger are used on the label.

In the case where some component or product of the compound is liable to the danger symbol T, the symbol C, X_i or X_n for another component can be left out. Correspondingly, the danger symbol C renders X_i or X_n unnecessary. Depending on the concentrations of the compounds of the dangerous substances, deviations from national regulations may be permitted.

[1] These risks and safety advice phrases are indicative. Risk and safety advice phrases applicable to the approximately 1,000 substances (and many times more compounds) required for labelling should be included in national regulations.

[2] The danger symbols used in the EC are illustrated in figure 52 of this manual.

If on account of the presence of various components, the compound or the product is liable to R-phrases of similar contents, these can be combined and a phrase indicating a lesser danger can be left out. The same goes for S-phrases.

However, if a danger marking is essential for safety, it should not be left out.

3.3. Size

The minimum size of the safety label on the packages depends on the volume of the package as follows:

Volume of package (V)	*Minimum size of safety label*
V < 0.5 l	A 9 (37 mm × 52 mm)
0.5 l < V < 1 l	A 8 (52 mm × 74 mm)
1 l < V < 10 l	A 7 (74 mm × 105 mm)
10 l < V < 50 l	A 6 (105 mm × 148 mm)
50 l < V	A 5 (148 mm × 210 mm)

EC risk phrases and their combinations

R 1	Explosive when dry
R 2	Risk of explosion by shock, friction, fire or other sources of ignition
R 3	Extreme risk of explosion by shock, friction, fire or other sources of ignition
R 4	Forms very sensitive explosive metallic compounds
R 5	Heating may cause an explosion
R 6	Explosive with or without contact with air
R 7	May cause fire
R 8	Contact with combustible material may cause fire
R 9	Explosive when mixed with combustible material
R 10	Flammable
R 11	Highly flammable
R 12	Extremely flammable
R 13	Extremely flammable liquefied gas
R 14	Reacts violently with water
R 15	Contact with water liberates highly flammable gases
R 16	Explosive when mixed with oxidizing substances
R 17	Spontaneously flammable in air
R 18	In use, may form flammable/explosive vapour-air mixture
R 19	May form explosive peroxides
R 20	Harmful by inhalation
R 21	Harmful in contact with skin
R 22	Harmful if swallowed
R 23	Toxic by inhalation
R 24	Toxic in contact with skin
R 25	Toxic if swallowed
R 26	Very toxic by inhalation
R 27	Very toxic in contact with skin
R 28	Very toxic if swallowed
R 29	Contact with water liberates toxic gas
R 30	Can become highly flammable in use
R 31	Contact with acids liberates toxic gas
R 32	Contact with acids liberates very toxic gas
R 33	Danger of cumulative effects
R 34	Causes burns
R 35	Causes severe burns
R 36	Irritating to eyes
R 37	Irritating to respiratory system
R 38	Irritating to skin
R 39	Danger of very serious irreversible effects
R 40	Possible risks of irreversible effects
R 41	Risk of serious damage to eyes
R 42	May cause sensitization by inhalation
R 43	May cause sensitization by skin contact
R 44	Risk of explosion if heated under confinement
R 45	May cause cancer
R 46	May cause heritable genetic damage
R 47	May cause birth defects
R 48	Danger of serious damage to health by prolonged exposure

Combination of R-phrases

R 14/15	Reacts violently with water liberating highly flammable gases
R 15/29	Contact with water liberates toxic, highly flammable gas
R 20/21	Harmful by inhalation and in contact with skin
R 21/22	Harmful in contact with skin and if swallowed
R 20/22	Harmful by inhalation and if swallowed
R 20/21/22	Harmful by inhalation, in contact with skin and if swallowed
R 23/24	Toxic by inhalation and in contact with skin
R 24/25	Toxic in contact with skin and if swallowed
R 23/25	Toxic by inhalation and if swallowed
R 23/24/25	Toxic by inhalation, in contact with skin and if swallowed
R 26/27	Very toxic by inhalation and in contact with skin

R 27/28	Very toxic in contact with skin and if swallowed
R 26/28	Very toxic by inhalation and if swallowed
R 26/27/28	Very toxic by inhalation, in contact with skin and if swallowed
R 36/37	Irritating to eyes and respiratory system
R 37/38	Irritating to respiratory system and skin
R 36/38	Irritating to eyes and skin
R 36/37/38	Irritating to eyes, respiratory system and skin
R 42/43	May cause sensitization by inhalation and skin contact

EC safety advice phrases and their combinations

S 1	Keep locked up
S 2	Keep out of reach of children
S 3	Keep in a cool place
S 4	Keep away from living quarters
S 5	Keep contents under . . . (appropriate liquid to be specified by the manufacturer)
S 6	Keep under . . . (inert gas to be specified by the manufacturer)
S 7	Keep container tightly closed
S 8	Keep container dry
S 9	Keep container in a well-ventilated place
S 12	Do not keep container sealed
S 13	Keep away from food, drink and animal feedingstuffs
S 14	Keep away from . . . (incompatible materials to be indicated by the manufacturer)
S 15	Keep away from heat
S 16	Keep away from source of ignition – No smoking
S 17	Keep away from combustible material
S 18	Handle and open container with care
S 19	Do not keep container sealed
S 20	When using do not eat or drink
S 21	When using do not smoke
S 22	Do not breathe dust
S 23	Do not breathe gas/fumes/vapour/spray (appropriate wording to be specified by the manufacturer)
S 24	Avoid contact with skin
S 25	Avoid contact with eyes
S 26	In case of contact with eyes, rinse immediately with plenty of water and seek medical advice
S 27	Remove immediately all contaminated clothing
S 28	After contact with skin, wash immediately with plenty of . . . (to be specified by the manufacturer)
S 29	Do not empty into drains
S 30	Never add water to this product
S 33	Take precautionary measures against static discharges
S 34	Avoid shock and friction
S 35	This material and its container must be disposed of in a safe way
S 36	Wear suitable protective clothing
S 37	Wear suitable gloves
S 38	In case of insufficient ventilation, wear suitable respiratory equipment
S 39	Wear eye/face protection
S 40	To clean the floor and all objects contaminated by this material use . . . (to be specified by the manufacturer)
S 41	In case of fire and/or explosion do not breathe fumes
S 42	During fumigation/spraying, wear suitable respiratory equipment (appropriate wording to be specified by the manufacturer)
S 43	In case of fire, use . . . (indicate in the space the precise type of fire-fighting equipment. If water increases the risk, add – Never use water)
S 44	If you feel unwell, seek medical advice (show the label where possible)
S 45	In case of accident or if you feel unwell, seek medical advice immediately (show the label where possible)
S 46	If swallowed, seek medical advice immediately and show this container or label
S 47	Keep at temperature not exceeding . . . ° C (to be specified by the manufacturer)
S 48	Keep wetted with . . . (appropriate material to be specified by the manufacturer)
S 49	Keep only in the original container
S 50	Do not mix with . . . (to be specified by the manufacturer)
S 51	Use only in well-ventilated areas
S 52	Not recommended for interior use on large surface areas
S 53	Avoid exposure – Obtain special instructions before use

Combination of S-phrases

S 1/2	Keep locked up and out of reach of children
S 3/9	Keep in a cool, well-ventilated place

S 3/7/9	Keep container tightly closed in a cool, well-ventilated place
S 3/14	Keep in a cool place away from . . . (incompatible materials to be indicated by the manufacturer)
S 3/9/14	Keep in a cool, well-ventilated place away from . . . (incompatible materials to be indicated by the manufacturer)
S 3/9/49	Keep only in the original container in a cool, well-ventilated place
S 3/9/14/49	Keep only in the original container in a cool, well-ventilated place away from . . . (incompatible materials to be indicated by the manufacturer)
S 7/9	Keep container tightly closed and in a well-ventilated place
S 7/8	Keep container tightly closed and dry
S 20/21	When using do not eat, drink or smoke
S 24/25	Avoid contact with skin and eyes
S 36/37	Wear suitable protective clothing and gloves
S 36/39	Wear suitable protective clothing and eye/face protection
S 37/39	Wear suitable gloves and eye/face protection
S 36/37/39	Wear suitable protective clothing, gloves and eye/face protection
S 47/49	Keep only in the original container at temperature not exceeding . . . ° C (to be specified by the manufacturer)